AF394699

TARKA
REVISITED

TARKA REVISITED

100 Years of Rivers and Wildlife

IAN PARSONS

PELAGIC PUBLISHING

First published in 2026 by
Pelagic Publishing
20–22 Wenlock Road
London N1 7GU, UK

www.pelagicpublishing.com

Tarka Revisited: 100 Years of Rivers and Wildlife

https://doi.org/10.53061/WACZ6902

A CIP record for this book is available from the British Library

EU Authorised Representative: Easy Access System Europe –
Mustamae tee 50, 10621 Tallinn, Estonia, gpsr.requests@easproject.com

ISBN 978-1-78427-629-4 Hbk
ISBN 978-1-78427-630-0 ePub
ISBN 978-1-78427-631-7 PDF

Cover artwork by Rachel Hudson

Typeset in Minion Pro by S4Carlisle Publishing Services, Chennai, India

5 4 3 2 1

Printed and bound by Short Run Press Ltd, Exeter EX2 7LW

Contents

Prologue

Fleeting glimpses, miniscule moments in time, a sinuous shape slipping smoothly below the surface, a disappearing dash into a culvert, a shadow amongst shadows… I had seen Otters, but conversely, I had never *really* seen them, their habitual elusiveness meaning sightings were brief. The word 'Otter' barely had time to pop up in my consciousness as my eyes registered these marvellous mustelid moments, and then they were gone.

Then one day, while pottering around on a riverbank – a bit of aimless mid-afternoon meandering on my part – I suddenly became aware of ripples spreading out across the water's surface, just a few metres away from where I stood in the shade of a bankside Ash tree. It was clear that something fairly large had just entered the water, but, puzzled by the fact that I had neither heard nor seen movement, I couldn't think what it could be. I stood transfixed by the ephemeral rings spreading out across the dark glassy surface in front of me.

I can still remember clearly the small bubbles rising to the surface, one after the other, creating a straight line of escaping exhalation, a line that betrayed the route of its source. As I watched the bubbles surface, so too thoughts began to surface in my head. I began to recall reading about the significance of straight bubble lines somewhere in the past. Slowly my brain kicked into gear, shedding its soporific slumber. Electrical signals flashed across my synapses, and suddenly I was alert to what was happening – my brain was telling me that those bubbles could be evidence of one thing, one quite amazing thing.

As the word 'Otter' broke into my thoughts, so the actual Otter broke the surface of the water; a sleek head popped up and looked straight back at me. It was a hard stare from this undeniably beautiful mammal, its dark eyes locked firmly on me, reading me, processing me. I stared back, my own eyes full of joy, as the Otter tried to assess my threat level, my meaning for it.

I am not sure how long our eyes locked, how long the optical connection between us lasted. It was probably no more than just a few seconds, but to me it felt like forever. It was the Otter that blinked first to end our mutual staring contest – a blink and then a nonchalant look around, dismissing my relevance, as it decided what to do next. It slunk back below the surface for a moment, before its head reappeared once more, pushing through the languid waters, creating a miniature bow wave as it swam slowly away from where I still stood, transfixed. The riverine mammal had a brief look back at me, and then it dived. My last glimpse was of the rudder-like tail, sleek in its coat of river, following the rest of the Otter below the surface. It was gone. It took me a few moments to realise that I ought to start breathing again.

Introduction

The year 2027 is the one hundredth anniversary of the publication of Henry Williamson's book *Tarka the Otter: His Joyful Water-life and Death in the Country of the Two Rivers*. It should be said straight off: I am not a fan of *Tarka the Otter*, to give its more commonly known abridged title. Please don't get me wrong – it is a beautifully written book of vivid and wonderfully descriptive prose, enabling the reader to picture the scenes, to see the action. But it is, for me at least, disturbing to read. I knew the outcome of the story before I first read it: the Otter, Tarka, dies for human pleasure.

Williamson follows the story of a male Otter, Tarka, from the point at which his pregnant mother settles down to give birth to him and his two siblings. The book is in two parts, the first of these entitled 'The First Year'. Tarka's birth, his eyes opening, his hesitant attempts at swimming are all covered as he begins to venture out in the world. His young life, totally dependent on his ever-attentive mother, is well documented and well described as Tarka develops and grows up, gradually but inexorably heading for independence before his first year is over.

The second part of the book covers his second year, but its title comes with a hefty spoiler; it is called 'The Last Year'. It follows Tarka, a now adult male Otter, as he discovers the River Taw, carves out his niche, and finds his own territory and a mate. There is another presence in the book though, a constant presence and one that is mentioned only a couple of pages in, his name written by Williamson before Tarka is even born. It is a presence that will bring the all too brief life of

Tarka to an end. This is an Otterhound called Deadlock, one of many hounds that make up the pack of the book's local Otter hunt. Otter hunting was not a fictional pastime when Tarka was written; it was a real and ever-present threat to Otters, and one that the author knew well – for Williamson followed his local hunt, and even met his future wife whilst doing so. His world-famous Otter book was dedicated to the master of that Otter hunt.

The barbarity of humans towards wildlife was ever-present in the era in which the book was written. It still is in some places, even today. And whilst we should never shy away from the reality of how wildlife is treated, I do find it difficult to read about in fiction. I read fictional stories for pleasure, and whilst *Tarka the Otter* is full of wonderful passages that paint beautiful pictures of an Otter's life, any pleasure gained is lost for me in the somewhat morbid obsession the author seems to have had with the various ways in which humans killed them. Gin traps, shotguns, a heavy crowbar – they are all utilised in the book's unremitting assault on these beautiful mustelids, but above all the danger comes from the dogs, the baying Otterhounds and their ripping jaws.

Like the fictional Tarka, I am from Devon, born and bred in the county where the author Henry Williamson also lived and died. His death came at the age of 81, when I was just four years old; I never met him of course, but even if I had, I am not sure I would have liked him. From what I have read of him, his views on the world are not just different to mine, but diametrically opposed to them. Fascist extremism is not something I can feel comfortable with.

Although I personally find the book an uncomfortable read, I cannot deny that it has had a massive resonance over the last hundred years. It has been, and continues to be, phenomenally successful. Since 1927, the year in which it was first published, it has never once been out of print. Its total sales are staggering, enough to make any author's eyes glow with envy.

One of the most powerful books I have ever read, and one of the most important works that everybody should read, is *Silent Spring* by Rachel Carson.[1] If any book is responsible for kickstarting the current environmental movement then it is this one, a powerful and compelling case against the development of pesticides and their widespread use. As you will see in this book, the use of these chemicals in agriculture and our everyday lives following the Second World War had a widespread and deadly impact on our wildlife, and the real Tarkas that inhabited our rivers were decimated by them. Rachel Carson's book created an awareness that ultimately led to the banning of many of these human-made deadly chemicals; if it ever could be said that a book saved the Otter from extinction in Britain, then that book would be *Silent Spring*. As we will see, however, that raised awareness of the dangers of these new chemicals took a long time to translate into awareness of how they were affecting animals like the Otter.

But there is also a case for saying *Tarka the Otter* played a role too – after all, Rachel Carson herself stated that the book had deeply influenced her as she grew up. Importantly, the book also brought these secretive, hard to see animals to life for its many, many readers; people who had little chance of encountering an Otter for themselves suddenly had access to their lives (and deaths). Whether it was intended or not, *Tarka the Otter* made people think about these riverine mammals like never before.

Tarka the Otter was set in an area of Mid and North Devon that the author Henry Williamson named Two Rivers Country. Those two rivers were the Torridge and the Taw, and in the second part of his book, Williamson narrates the tale of Tarka finding the source of the Taw and then following its flow all the way to the sea. I write this sat at my desk just a few hundred yards away from the Yeo, a major tributary of the Taw; the Taw itself is less than four miles away, and is a river I know well. It is a beautiful watercourse, starting on the high ground of

Dartmoor and spilling into the sea at Bideford Bay in North Devon.

I want to take you on a journey of sorts along the River Taw. From its bubbling beginnings to its estuarine end, we will in effect be following the fictional Tarka a century after he made the journey. But this metaphorical journey down the Taw is not just along this particular river; it is also a journey through a hundred years of Otters in Britain, exploring the river systems in which they live and the other wildlife with which they share those rivers. A lot can happen in a hundred years.

CHAPTER 1

Bubbling Beginnings

I look up as I walk along the broad, empty stone track. Above me there is a bright blue sky punctuated by towering dots of Skylarks, their liquid song tumbling back down towards the moor I am traversing. The song of the laverock is pouring down all around me, soaking me in its aural beauty.[1] Other bird sounds accompany me as I walk through the unceasing openness of the moor, but none are as musically pleasing as the larks, for they are the harsh scolds made by resident Stonechats and visiting Wheatears, objecting to my presence on the land they are declaring as their own. The path that I follow starts the inevitable climb upwards, the incline tugging at my leg muscles as I slowly ascend the gradient. I look up with envy at a Common Buzzard effortlessly gaining height, spiralling itself into the blue sky on broad unflapping wings. I keep walking.

All rivers start somewhere; their beginnings are typically very humble, a gathering of drips seeping out of the ground, falling together before trickling onwards. The River Taw is no exception. But whilst the first clear-water drips of the Taw might be modest, the setting for their emergence from the sodden ground is not – it is glorious. The river rises at a place called Taw Head on the high ground of North Dartmoor, a vast expanse of open moorland that feels empty of humans and full of wildness.[2] Not all rivers originate in such dramatic landscapes, and not all rivers flow through National Parks.

The rivers of this country are as varied as they are many, but they have all been through the last hundred years.

Taw Head, when I eventually reach it, is unsurprisingly a wet place. The immediate area around it, a site that lies to the west of the imposing Hangingstone Hill, is the source for other Devon rivers such as the Teign, the Okements (which later converge and flow into the Torridge, the other river in Williamson's Two Rivers Country) and the Dart,[3] which rises no more than a few hundred metres away from the Taw. Those few hundred metres make the world of difference in terms of where these two rivers flow, with the Dart eventually disgorging itself into the English Channel in South Devon and the Taw debouching into the Bristol Channel in the north of the county.

It doesn't take long for the Taw's trickle of drips to coalesce into a small bubbling stream. A mere handful of metres is all that it takes for a patch of wetter-than-normal land to become a narrow flow cutting through the thin moorland turf. The river's inevitable journey to the sea has begun.

But before we accompany the river on its journey, we should first get to know the main protagonist of both this and Williamson's book.

What is an Otter?

Our Otter, scientifically known as *Lutra lutra*, has the formal title of Eurasian Otter. It is one of thirteen species of otter in the world, but it is the only one that occurs naturally in Europe. However, our Otter isn't solely a European species, for this animal has a vast natural range. It is found from the Atlantic coasts of Portugal and Ireland in the west, right through to the tropical coasts of South East Asia in the east; and from the Arctic seas off north Norway to the Atlas Mountains of North Africa. The 'European' Otter is a truly international species.

The Otter is a mustelid, a member of the Mustelidae family that is found within the large mammalian order of the carnivores – an order that also contains the dog, cat and bear families, amongst several others. The mustelids are often informally referred to as the weasel family, and indeed the small but very formidable weasel is the archetypal mustelid, the classic example of a family that tends to follow the formula of short legs, an elongated tube-like body and powerful jaws.[4] The name mustelid refers to this body shape; it is taken from Latin and means a 'mouse-shaped spear'.

The Otter is in effect a semi-aquatic weasel, albeit a fairly large one when compared to the other members of the mustelid family that inhabit our shores here in Britain.[5] Like many members of the family, Otters exhibit sexual dimorphism, with the males being bigger than the females; a male Otter can attain a length of 1.2 metres and weigh in at around 8 to 10 kilograms. The smaller and more slender females reach lengths of around a metre and weigh in at between 6 and 8 kilograms. They are predators, mainly catching and eating fish, but they will also ingest crustaceans such as crayfish and crabs, as well as the odd amphibian and, less frequently, small mammals and birds.

Their predominantly fishy food makes them piscivorous, literally fish eaters, and studies have shown that this accounts for up to 90% of their diet. But what sort of fish – are they generalists or do they have preferences? The answer is a bit of both: in freshwater river systems Otters tend to feed on the fish that are readily available, be they Sticklebacks or Salmon, but where there is a good selection of prey items, Otters tend to favour certain species – generally the ones that are easier to catch, with Eels being a particular favourite. In coastal sea waters, their prey tend to be slow-moving, bottom-dwelling species found close to the shore, a habit which makes them relatively easy for the Otter to catch; other species that swim in the deeper open-sea waters rarely feature in their diet.

In both fresh and salt water, therefore, how catchable a fish is tends to determine whether it features in the Otter's diet. But how do Otters catch their fish? In clear water they will use their eyesight to locate and then pursue fish; in deep water Otters have been reported as taking prey from below, with the fish being easier to spot against a background of sky rather than the bottom of a watercourse. But rivers and watercourses aren't always crystal clear – they are often turbid, murky places, and when there is disturbance in the water such as waves or eddies and white water, visibility lessens considerably.

If you are ever lucky enough to get a close-up view of an Otter's face, you will notice a significant array of whiskers on its muzzle. It is these vibrissae, to give them their 'proper' name, that are key to the Otter's hunting ability in visually obscured waters. The importance of these whiskers was demonstrated by a researcher studying two tame captive Otters in a large pool of water. The researcher timed how long it took the Otters to catch fish in clear water, before timing them again after the water had been heavily obscured by adding a large quantity of powdered charcoal to it (the water was so murky that a human observer could only detect a shiny metal object in it if it was within 10 cm of the surface).

Unsurprisingly, the Otters took longer to catch the fish in this second part of the experiment – on average, four times longer. The researcher then trimmed back the whiskers of one of the Otters, effectively removing them (they would grow back) and the benefit they gave the animal. In the clear water, the trimmed-back Otter was able to catch fish as quickly as before, demonstrating that eyesight was indeed the primary sense used. In the murky charcoal-laden waters, however, the Otter took twenty times longer than it had pre-trim. The loss of the whiskers clearly impeded its hunting ability in low-visibility conditions.

But how do these whiskers help the Otter locate its prey? Do they physically touch the fish, or do they somehow detect its presence? We use the term 'whiskers' loosely as humans, often referring to our own facial hair, but animal whiskers are not the same thing – they are specialised sense organs that can be found on many mammals. The whiskers on an Otter's muzzle aren't just specialised, but highly specialised, far more so than those of our pet dogs and cats, and more so even than their close mustelid cousins. Under microscopic analysis, the vibrissae of an Otter are more akin to those of a seal than they are to a weasel's; but that shouldn't come as a surprise, for seals, just like Otters, hunt fish in water that is often far from being crystal clear.

Hydrodynamic trail following is hardly a phrase that trips off the tongue, but it is something that both seals and Otters do. When a fish moves in the water it creates turbulence within it; if that fish is swimming away from the Otter or seal it is leaving a trail of this turbulence, a trail that is stronger closer to the fish and weaker further away. This is a hydrodynamic trail and if followed, it will lead the follower to the cause. The vibrissae detect this turbulence – they detect the trail of movement made by a fish, even though the fish itself cannot be seen. This is how Otters are able to pursue and catch their prey even in extremely turbid waters. Fish can remain still in water and when still they are not creating turbulence, there is no hydrodynamic trail to follow, but they can still be vulnerable. A large mammal like an Otter diving and swimming through water is going to be very noticeable to a fish. If its nerve holds and it remains still, it may indeed go undetected – but if the presence of the Otter spooks it, if the fish, in effect, bottles it and moves, then that trail is created and the Otter can detect and follow it.

In flowing water, fish can remain in one position in the current, but to do so they have to be moving. If you look down from a bridge at a flowing river and see a fish below you that appears to be stationary, it is in fact swimming, matching its swimming speed to the speed of the current. The turbulence created by the fish that is in effect treading water might be slight, but it is still detectable to an Otter. That is a remarkable ability, but it is made even more remarkable when you think of a flowing river and turbulence.

Moving water engenders turbulence. Riverbeds are not smooth places – all the stones, rocks, silt banks, even shopping trolleys that are found on them create turbulence, as will the banks, protrusions in the water such as boulders, sudden drops or rises in the riverbed, and so on. What makes the remarkable ability of hydrodynamic trail following even more extraordinary is that the Otter is somehow able to differentiate the turbulence that is created by a fish from the myriad other causes of turbulence within a pretty chaotic environment. We humans love to dream up characters that have super-senses for our stories and films. Nature does it for real.

But the whiskers are also used for touch too, especially when the animal is exploring the riverbed and nooks and crannies within its territory. Because Otters are territorial, they get to know where prey items tend to hide; if an Otter has found food once, hiding in amongst a mass of submerged roots on a riverbank, it is highly likely to check that site again in the future. When feeding in this manner the Otter will literally stick its nose in, pushing its muzzle and the abundant array of vibrissae into any gaps in the cover to detect by touch anything edible lurking there. It will also use this method when turning over submerged stones. Even in clear water, turning over a stone is going to kick up a cloud of obscuring debris, but the whiskers can help detect if there is anything hiding underneath. It is with this method that the Otter hunts for crayfish, which will often lurk under cover on the riverbed. Coastal otters also employ a similar mode of hunting when searching in and under seaweed, especially in large rock pools where they are able to find crabs.

The Common Frog and Common Toad also occur in the diet of Otters, but tend to do so seasonally, either when the frogs are hibernating in the mud at the bottom of ponds, when the whiskers would be employed to find them, or when the frogs and toads are spawning and large numbers of them accumulate on the edge of water or in the shallows. When the amphibians are spawning the Otter has no need for super-senses; it can quite simply walk amongst them and eat at will, helping itself as it sees fit – a *Bufo* buffet in the case of toads, or a frog feast.

Whilst they are skilled hunters, Otters are also opportunists, and it is this that probably accounts for the appearance of mammals and birds in their diet. There is some evidence that Otters will catch waterfowl by swimming underneath them and then grabbing them, but the likelihood is that most birds that turn up in the diet are taken from ground nests that the Otter has come across, or from birds that are roosting and have been surprised by the Otter. Otters don't actively hunt small mammals such as voles and mice, something that the smaller mustelids such as Weasel and Stoat do, but if an opportunity presented itself the Otter would, like any good predator, readily take advantage of it.

However, it does appear that Otters will actively hunt Rabbits, a good-sized mammal that would provide a decent meal. Rabbits are not native to Britain – they were introduced by the Normans at some point after the 1066 conquest[6] – but they swiftly became a fixture of the landscape, and the opportunistic Otter must have soon learnt how to hunt them. Otters are not built to chase down fast mammalian prey, so don't expect to see one chasing a Rabbit across an open field, but they are built like most other mustelids in that they are long and slender and have short limbs. In other words, they are well shaped for entering burrows, especially Rabbit ones in loose ground where the holes and tunnels tend to be bigger, exactly the sort of Rabbit burrows found on the alluvial soils alongside river systems.

There is a possibly apocryphal tale of a researcher who was tracking a radio-collared Otter over a period of several nights. Each night the Otter's signal would drop off in exactly the same place for several minutes at a time before magically appearing again, causing mystery and consternation for the researcher in equal measure. It was only during daylight, when the researcher explored the area where the signal always dropped off, that the reason was discovered – a large Rabbit warren. The Otter was making nightly visits to it and when it slipped into the warren, heading underground, the radio signal would fail.

Whilst the veracity of that tale may be questionable, Rabbits do indeed form part of the diet of Otters, and back when Rabbit numbers were much higher in Britain than they are today, they were probably of more importance in the Otter's diet.[7]

Despite their predilection for Rabbits, Otters are without doubt primarily fish eaters, and to eat fish they have to enter the water. The waters of Britain's rivers are cold, so cold that warm-blooded mammals such as us will rapidly lose body heat in them. This in turn can soon render them dangerous places for us to dwell in, especially in the colder months of the year. Because of this, one would therefore expect any mammal that spends a lot of time immersed in their cold waters to have a layer of fat to protect their inner workings from the warmth-sapping water.

The archetypal aquatic mammal is perhaps the seal, and there are thirty-four species within the pinnipeds, or seal family – all of which are of course found in water, often very cold water. The pinnipeds are actually relatives of the Otter and the other mustelids, being more closely related to them than any other mammal family.[8] Seals are well known for their blubber, a layer of fat between the skin and the muscles of the animal. It is this blubber that insulates them from the cold waters in which they dwell, and it can be a substantial layer of fat, with blubber making up as much as 50% of the body weight of some seal species.

But the European Otter does not have an insulating layer of blubber; in fact, it has hardly any body fat at all. Otters are very lean machines. Typically, just 3% of an Otter's body weight is fat.

To put that into perspective, for humans a healthy body fat percentage is somewhere between 10% and 30%.[9] Otters have no layer of blubber to keep them warm whilst swimming in cold rivers, and they have hardly any fat to insulate them on those frosty mid-winter nights – but they do have a fur coat, and what a fur coat it is.

Coats and courtship

The European Otter's cousin, the Sea Otter of the Pacific Coast of North America, has the thickest fur of any mammal, with up to an astonishing 150,000 individual hairs per square centimetre. The fur of our European Otter doesn't come close to that. It has up to 80,000 hairs per square centimetre, but this is still an incredible amount, especially when compared to us. Even the most hirsute head of hair on a human contains no more than 300 hairs per square centimetre, so a figure of 80,000 is almost unimaginable to us.

The fur of an Otter is actually made up of two different types of hair, the first being the guard hairs, which are waterproof and keep the moisture out of the second type, found beneath them, the very dense underfur. It is this underfur that keeps the Otter warm, trapping air within it, air that acts as an effective insulator against the cold temperature of the water. But for this underfur to be effective it must remain dry at all times, even when the animal is underwater. The guard hairs accomplish this by keeping the underfur dry, but they cannot do it indefinitely – which is why an Otter will spend a lot of time out of the water, grooming. It has to keep its furry coat in tip-top condition in order to survive.

When we think of Otters, we generally think of them as being in the water, imagining them swimming and diving, but in fact they spend very little time actually submerged. They spend most of their time in their holts, the name given to the den of an Otter, in which they rest and sleep. But even when they are active and out of their holts, they don't spend as much time in the water as we are inclined to think. Studies have shown that an Otter will typically spend fifteen minutes actively hunting in water before coming out onto land (or back to their holt) to rest and groom for a similar length of time. Their thick fur coat is a great insulator, but being immersed in cold water soon reduces its insulating properties and cools them down. Taking a break from aquatic action not only gives them a rest, but more importantly it also allows them to warm up and to maintain their guard hairs to keep them clean and waterproof. The grooming process also restores the air, that all-important insulator, to their underfur. Otters need the water to hunt in, but they also need the land to provide them with secure resting places in which they can groom and holts in which they can reside and breed.

Adult Otters are solitary in nature; they are highly territorial animals. For the vast majority of the time males and females avoid each other wherever possible, living their separate lives, but for a period of around just five days, when a female Otter comes into breeding condition, male and female Otters will spend all their time together. Their courtship at first involves lots of play and mock fighting as the two animals assess and get to know each other. They will then mate several times during the five or so days they are together, with mating typically occurring in the water. But once this brief period of courtship is over, once the female has conceived and is no longer in season, the two adults separate and become solitary once more. The male European Otter plays no role in the subsequent rearing of any young resulting from their brief liaison.

Many animals have a breeding season, but the Otter in Britain generally doesn't and young can be produced in any month of the year. This is likely a result of not only our relatively mild climate, but also the year-round presence of sufficient food. In the parts of their range where winters are much more severe, such as in Scandinavia, there is a clear breeding season with cubs being born in the spring. This ensures that when a female is rearing her cubs she is able to easily find enough food to sustain her and her offspring – something that would be extremely difficult, if not impossible, in the Scandinavian winters when rivers and lakes are routinely frozen and food availability is therefore limited. A breeding season is apparent in the very north of the British Isles in the Shetland Islands, where the majority of young are born in the summer months. It is tempting to think this is due to winter weather but it is actually due to food availability again, with far fewer fish being present in the coastal inshore areas, where the Otters are able to hunt for them, during the winter months.

The gestation period of a European Otter is about 63 days or nine weeks; the number of cubs in a litter varies between one and five, but typically in Britain it is two or three. The young are born furred and measure about 12 cm long; they are helpless and depend totally on their mother. After about two and a half weeks they are able to crawl around the confines of the holt, and their eyes open after four to five weeks. The female suckles the young and the young get all their food from her for the first seven weeks of their life, after which point the youngster's teeth start to erupt. At this stage they are introduced to solid foods, initially small pieces of fish that the female has chewed up for them, and the proportion of solid food in their diet steadily increases until by fourteen weeks of age the young are fully weaned.

Once the young start feeding on solid food, they become increasingly active, venturing out of the breeding holt onto land and exploring their immediate surroundings. It is not until they reach about three months of age, however, that they make their first venture into water. I don't know what a young Otter makes of the first time it enters the water; one would imagine it being extremely nervous and perhaps even reticent to do so. But however it feels about it initially, it soon becomes second nature for the young animal to submerge itself and swim about – perhaps not with the full gracefulness of an adult at first, but it soon becomes at one with its new-found environment.

I have had the privilege of watching a wild family of Otters, a female and two young, over a period of a few days in central Spain, the female being constantly badgered (if that is the right word!) by the youngsters as she swam through the shallow waters and trotted across the banks and shale islands in the slow-flowing low waters of the large river by our village. The youngsters' behaviour was a mixture of supplication, brave exploration, nervous reaction and quite often foolhardy play, whilst the female was continuously alert to possible danger, as well as to the opportunity of finding some food to satisfy the youngsters' begging. The act of finding food for herself and her young came with two main issues: firstly, stopping the youngsters from getting in the way of her hunting. I can recall one scene in particular where the female had clearly identified that something edible was beneath a large stone on the water's edge (I suspect it was a crayfish, which are particularly abundant on this river). Just as she managed the Herculean effort of flipping the stone, the two cubs jumped on her from the side, sending her sprawling, before leaping on top of their mother and preventing her from getting back up again, allowing whatever was under the stone to slope off unnoticed. Secondly, when she was able to hunt for food unimpeded, she had to also manage the dilemma of the unattended cubs and their safety from not just potential predators but also themselves.

On several occasions I watched one of the cubs, entranced by its own curiosity, wander away along the watercourse whilst the female was trying to procure food in one of the large areas of deeper water. The cub's confidence seemed supremely high as it got further and further away from the female, before something would startle it out of its daring exploration and it would suddenly realise it could no longer see its mother and was all alone in a big world. This invariably led to a state of panic in the youngster and an inability to retrace its steps along the linear feature it had wandered. Its answer to this panic was a series of loud whistle-like calls of piercing plaintive character, a call that was quickly answered by the mother and the other cub that had stayed closer to her. On hearing this reply, the lost youngster would immediately retrace its steps with a slightly clumsy gait, calling all the time until it came within sight of the female, at which point it would increase its speed, running to her before bounding all around and over her. At which point the second cub would join them and all three would partake in what appeared to be a bout of reassuring grooming and nuzzling. Before long the three of them would continue to explore together once more, the youngsters, their confidence somewhat shorn, staying close to their mother – but, after about two minutes of this, one of the youngsters would get distracted by something and it would happen all over again… Female Otters rearing young could probably do with some therapy.

The young Otters stay with their mother for many months, initially relying on her for all their food, but as each day passes, they will find opportunities to hone their own skills, gradually learning how to catch their own food. After about nine months, the young Otters start to disperse, although some will stay with their mother for longer, sometimes for as long as twelve months. Even those that are reluctant to sever the apron strings, however, have to do so in the end, and when they do disperse they are entering an uncertain and dangerous period of their lives.

A newly dispersed Otter finds itself alone and inexperienced. It will also find itself most unwelcome in another Otter's territory, a state that can escalate to aggression should the young Otter linger. Sub-adult Otters usually end up having to occupy suboptimal habitat, areas where food and cover are unlikely to be prevalent. As a result they become somewhat nomadic, travelling long distances trying to find a vacant space to call their own. The second year of an Otter's life is a very hard one, and it is no surprise that those in their second year suffer a far higher rate of mortality than in subsequent years.

Otters are sexually mature from two years of age, around the time they are likely to have become settled in their own territory, replete with good food availability and potential holt sites. The cycle then begins again.

An invisible king

Most people are familiar with Otters – after all, their image is used to grace a wide variety of products, everything from beer to garden centres, and most people will be able to recognise a photo of one. In reality, however, people very rarely actually see one. As we will explore, the Otter is an animal that is now doing relatively well in Britain – they are to be found right across the whole of the country, from quiet rural rivers to busy urban ones – but they remain typically elusive and on the viewing wish list for many of us who are interested in wildlife. There is a particular aura about seeing an Otter, partly down to the fact that they are such beautiful creatures, but also because they are so secretive in their habits. The writer Roger Deakin summed it up perfectly in his book *Waterlog*: 'the animal's particular magic does not stem so much from its rarity as its invisibility'.[10] To go looking for Otters can be, and usually is, a frustrating experience, but just because you don't see them doesn't mean that they haven't been there.

At first glance the beginnings of the Taw, as with most river sources, does not look like a place you would expect to see Otters. There is barely enough water for the mammal to paddle in, let alone swim, and there is very little cover on the land for these mammals to safely groom and rest in. I often spend time on the upper elevations of Dartmoor; it is a wonderfully remote place and when I stand by the nascent River Taw, high up on the moor, I certainly don't expect to see an Otter. It is not a place where I would come looking for one, but that doesn't mean that an Otter has never passed through, never splashed in the shallow waters. As we have just outlined, when an Otter first becomes independent from its mother, it becomes something of a watercourse nomad. Trying to find a place they can call home along the linear riverine habitats in which they roam is not easy for a young sub-adult Otter, especially when there is an already well-established, and very territorial, Otter population present. These young wandering Otters can and do travel long distances, following the watercourses upstream and downstream, in an attempt to find a place in which they can safely settle.

Travelling upstream will inevitably lead you to a river's source; for an animal travelling upstream, the beginning of a river is also its end, but for a young Otter it is not the end of the journey. We typically tend to think of Otters as travelling by river, dispersing themselves via water, but they will also cross large areas of dry land if required. Their arched-backed gait and short legs gives them an ungainly appearance as they move across the earth, but still they are perfectly able to do so, and such movements allow them to cross from one watershed to another, switching between river systems. For an Otter to leave the watershed of the Torridge and enter that of the Taw, as the fictional Tarka did, it would only need to cross about 400 metres of moorland from the source of the West Okement, to reach the source of the Taw. As I stand by the Taw's small stream on the highland of the moor, I can picture an Otter doing just that. It is easy to imagine one slinking over the undulations, using the granite rocks as cover, heading for a new river system to explore.

As we have seen already, for most of their lives European Otters are solitary animals, living within their own territories and eschewing social contact; these territories are marked, patrolled and defended, and new settlers are not permitted. It is this aspect to the lives of Otters that forces young, newly independent animals to wander as they do, trying to avoid conflict with more mature established animals, trying to find a place in which they can set up their own territory.

But any young Otter entering a river at its beginnings, especially a moorland river like the Taw, is unlikely to find a suitable spot for that territory. To do that it will have to travel downstream, following the fledgling watercourse as it develops into a more suitable riverine habitat that can provide them with both food and cover in which to rest. All rivers are different; some develop quickly, deepening and broadening out, whilst others are slower in their development, retaining their small stream-like qualities for longer. The Taw falls into the latter category, but you cannot really call it slow.

The wet ground of Taw Head coalesces into a moving, flowing watercourse at about 560 metres above sea level. This tiny shallow stream then meanders slowly around a memorial stone to the late Poet Laureate Ted Hughes, himself a friend of Tarka's creator and who aptly described an Otter in one of his poems as acting like 'a king in hiding', a phrase that reflects their majesty as well as their elusiveness.[11] The young Taw then stumbles rapidly down Steeperton Gorge, a long drawn-out set of continuous miniature rapids, meeting its first tributary, Steeperton Brook, at the point where the still-narrow watercourse plateaus out on to Taw Marsh. In its short existence, in its brief two and a quarter miles of flow, the young Taw drops 200 metres in altitude as it quite literally falls off the high land of North Dartmoor.

Taw Marsh is an area of wet, flat moorland lying between two ridges of higher ground through which the river snakes at a laconic pace; the river here is shallow and narrow as well as exposed, still lying at over 350 metres above sea level. There is little here to hold a wandering Otter's attention. It is only after the river exits the marsh, three and a half miles after it sprang from the sodden peat of Taw Head, that it starts to resemble a river an Otter might regularly use. The majority of rivers in Britain become suitable for Otters very quickly after first emerging from the ground – three and a half miles is an unusually long distance, but from this point on, the River Taw starts to become a river for Otters.

The pace of the river becomes varied; there are slow deep pools interspersed with shallow fast rapids. The river widens and the sides steepen, and a scattering of trees – Common Hawthorn and Rowan, protected from the incessant grazing of sheep by the steeper sides and the granite boulders – grow alongside the water. The roots of these trees spread across the ground, finding gaps into which they can grow downwards to grip the substrate. Here and there exposed granite rocks form untidy piles, the soil that once lay between them long washed away by winter floods and heavy rains. It is these rocks and the tree roots that more often than not provide the Otter with a place of shelter, a place to which it can retire to groom that all-important fur and rest. Only when a river starts to provide these sorts of places does it become a river for an Otter to spend time in, rather than one to merely travel through.

A place to call home

The den of an Otter is called a holt,[12] a term that was used by Otter hunters to describe a place where an Otter would lie, incorporating a tunnel leading to a chamber. In other words, a holt is a resting place with an entrance and a roof. This distinguishes it from another Otter resting site, a couch, which is an open-air place in which an Otter will lie, groom and rest. One classic example of an Otter's couch is a flattened platform of reeds within a reed bed.

Unsurprisingly, most holts are situated close to the water, with the majority located on and in the riverbank itself. The most popular sites are those formed by tree roots, the spreading network of roots forming a protective shelter over cavities underneath. These cavities are often caused by the water itself, but they are also formed when a tree is partially toppled by the wind, or when the tree's natural lean out over the water lifts its root plate to create a hole between it and the soil beneath. The granite rocks that litter the banks of the moorland Taw are a lithic version of tree roots over cavities, and where rocks like this occur, they will be readily used. Elsewhere, these naturally formed rock sites also have human-made versions; where boulders have been placed within the river to reinforce embankments and in flood defence schemes, these too will often be used. Otters are very adaptable when it comes to holt sites, and my favourite record of an Otter holt is one noted in Paul Chanin's book, *Otters* – it was the back seat of a car that had been abandoned alongside a river![13]

Other similar-sized mammals in the UK, such as the Fox and the Badger, excavate their own dens (or setts in Badger terminology) – that is, they physically create them. Otters, however, do not; they are entirely reliant on naturally occurring sites or, more recently, human-made ones that they can utilise. Male Otters aren't generally fussy about where they will lie up for a rest, but females looking to give birth need to have a secure site or sites in which to do so. If a river lacks secure sites, it will lack breeding Otters too. Canalised rivers as part of drainage projects and flood prevention schemes may be full of potential Otter food, but they offer very little in terms of holts and generally only contain visiting animals rather than resident ones.

Because Otters do not excavate their own holts, they have had to rely on the distribution of naturally occurring ones. Such sites are often limited in number and distribution within a river system, and as a result, a good holt site can be one that is used by generations of animals over many years. A good solid holt might be in use for decades and even centuries, a fact that was exploited to the full by Otter hunters. Once a holt became known to them, it became the first port of call when they sought their quarry. For young Otters seeking out a home for themselves in the era when Tarka was written, a vacant holt and territory may have seemed very opportune, but it was also, most likely, a double-edged sword.

I only have to walk a couple of hundred metres downstream from the point where the Taw begins to look a lot more ottery to find my first encounter. It's one that has the potential to tell me a lot about the Otter in question – its sex, its age, its breeding status – but alas, my olfactory skills are simply not up to the task. For my encounter is not with an actual Otter, but with its calling card, its droppings or spraint.

Just after the river has passed through a long dismantled concrete weir, a narrow line of granite forms an island in the middle of the widening watercourse. In places this small rocky ridge is overrun by the flow, awash with water, but in one of the drier parts a small clump of rush has found a foothold and is growing surrounded by the moving water. The centre of this rush clump has clearly been flattened, its stalk-like leaves splayed to all points of the compass, and at its flattened centre lies a spraint. I am unable to get a close look, as to do so risks wet feet, but from where I am stood the spraint looks quite dry, so doesn't appear to be fresh. After a good look around, I deduce that it also appears to be isolated, a lone sign, indicating to me that this part of the Taw is a bit of a backwater, the edge of a territory that only gets visited occasionally. Nonetheless it is proof that Otters are here, the first Otters of the Taw.

Such droppings are a lot more than just discarded waste matter; the spraint of an Otter is also a complete package of information. For these animals that mainly lead solitary lives, spraint is their primary method of communication, a noticeboard for other Otters to read and comprehend. For nature enthusiasts like myself, a spraint can often be the only sign that there are Otters about, a proof of their existence, the elusiveness of the animal betrayed. The word 'spraint', like the word 'holt' before it, derives from Otter hunting terminology, but is believed to originate from a French word that means 'to squeeze out'.

Otters have scent glands inside their rectums; these glands coat the faeces that pass through with a combination of pungent chemicals. Otters produce a lot of spraints and regularly replenish them – they are a significant part of an Otter's life. Exactly what these spraints tell other Otters is largely conjecture on our part, but they certainly signal that another Otter is present, or at least has been present recently, and it is believed that Otters are able to recognise individuals by their spraints. It is also widely thought that the spraint passes on a wealth of information regarding the individual: its sex, breeding status, and so on. A young Otter travelling through and exploring a new river system would soon know that a territory was occupied if it kept finding spraints produced by the same animal.

Our sense of smell is very poor when compared to other mammals, but even we can pick up the distinctive scent of an Otter spraint and differentiate it from other similar droppings like those of an American Mink. There are, and have long been, various human definitions of what an Otter spraint smells like, but unlike many other mammal droppings they do not have a repulsive scent. Instead they emit an almost sweet and spicy aroma, a bit like jasmine tea. Many years ago, a mammal enthusiast told me that if I smelt one and it made me want to be sick, then it was not an Otter's spraint. Maybe not the most specific of definitions, but it is one that works!

The spraint I spotted in the midst of the Taw wasn't accessible for me to smell, but I am confident in my conclusion that it was an Otter spraint. Its location obviously helped, on a small island in the middle of a fast-flowing river, but so did its appearance. An Otter's spraint is typically around 6 to 8 centimetres long and is cylindrical in shape, tapering to a point at one end. The one I saw was black in colour, but this is very variable and can be influenced by the weather, the age of the spraint, and what the Otter has been feeding on – the spraint of an Otter that has been gorging on crayfish can be pink in colour. Along with its general shape, what really helped me to identify it was the presence of numerous white fish bones dotting its surface, their distinctive spiny appearance a real giveaway.

Shortly after my encounter with the spraint, the River Taw passes the first human habitations, outliers of the small village of Belstone; and a few hundred metres after that, at the point where the river turns sharply to the east, it is crossed by a footbridge that leads to a number of paths and tracks. Here is a gateway to the high moor for many people, and the shallow gravelly pool that the river forms at this point is incredibly popular. It is the point where the Taw becomes extremely public, and whilst the vast majority of people that visit this site will be unaware that Otters use this part of the river, the Otters themselves will be very aware that humans use it.

CHAPTER 2

Us and Them

The patterns of the flow are mesmerising, smooth swirls interspersed with foaming turbulence. A piece of bracken, adrift in the current, spins slowly clockwise as it moves inevitably onwards. Then the gyration stops, its trajectory straightens, and the fragment of fern is suddenly picking up speed, shooting along towards some protruding lumps of granite – the water forced up the face of the lithic obstruction, forming a watery collar, before crashing downwards into a chaotic mix of movement. The torn bracken raft vanishes below the surface as it plunges past the igneous rock, resurfacing several metres downstream, travelling rapidly onwards until it reaches a stretch of calm. Here the surface of the water is smooth with the reflection of the sky, and the bracken arcs gently and serenely to the side, trapped in an eddy of tranquillity, joining an increasing flotilla of vegetative flotsam that the river has gathered in this spot.

The sound of the river lessens in volume, the crash of the turbulence replaced by the more soporific murmur of calm flow. This reduction in riverine noise allows my ears to engage with other sounds: a snatch of song from a Robin and then the sound of human voices, both coming from a garden. The young river is about to pass by its first human dwelling.

Humans and Otters have a long history, the majority of which is not favourable for the Otter. Humans have long hunted the animals for food and for fur, with Otter remains being found in the middens of human settlements from 6,000 years ago. Luckily for the Otter, we humans apparently were not all that taken with the taste of its flesh, which has been described in some historic accounts as simply rank – not exactly a recommendation. But, unluckily for the Otter, we were very much taken with its fur. The wonderfully thick and insulating fur described earlier was a highly valued commodity in the human world.

Tarka was not the only famous fictional character from Devon with an Otter fur coat. John Ridd, the hero of the novel *Lorna Doone* by R. D. Blackmore, proudly wore his favourite Otter-skin jerkin on many occasions throughout the book, and there is no doubt that the fur of an Otter would have made a highly prized item of clothing.[1] Hunting Otters for fur would have had localised impacts on populations, making them scarce in areas where such hunting was concentrated; and indeed records show that the trade of Otter pelts in fur markets in south-west Scotland rapidly fell off in the mid-1800s as the number of Otters available to hunt locally presumably declined because of their over-exploitation. But nationally, Otter hunting for fur in Britain was not a major cause of widespread population decline. It is interesting to note, though, that commercial Otter hunting for fur carried on in Britain (in the Hebrides) right up until the late 1970s, when Otters finally became legally protected.

More impactful on Otters and their numbers was their classification as vermin in Queen Elizabeth I's 1566 'Acte for the preservation of Grayne', a statute widely known today as the Vermin Act. This act decreed that you could be paid the princely sum of one penny for every head of Otter you were able to produce. The Otter was on the 'vermin' list alongside many of the usual victims of human prejudice – species such as Fox, Badger and most birds of prey – but there were a few surprising species listed as well. The Green Woodpecker was included because it had an apparent predilection for damaging the wooden shingled roofs of churches, whilst the Hedgehog made the unfortunate list due to an erroneous belief that it would steal the milk from cow's udders whilst the cattle slept at night. Nothing like a bit of rigorous scientific justification for our war on wildlife! The third surprising species on the list gives a clue as to why the Otter was there too: the Kingfisher. This magnificent jewel of our avifauna found itself on the list simply because it dared to eat fish. Kingfishers are diminutive birds, virtually sparrow-sized, and therefore any fish they catch are themselves very small. Research shows that on average, fish caught by Kingfishers are around 2.3 cm in length; you would have thought this hardly made them competitors to us humans in terms of fishing, but in the eyes of the Elizabethan legislators it most certainly did. If a Kingfisher can be declared as a pest due to its miniature piscivorous feasting, then it should come as no surprise that Otters were also persecuted for their own very fishy diet.

Freshwater fish formed an important part of the British diet during the medieval and post-medieval periods; they were a key food source for all sectors of society, from peasants trapping and netting fish in streams and rivers, through to large houses and monasteries that kept well-stocked fish ponds. These same fish were, of course, also a key food source for Otters. Whether you were landed gentry or landless serf, the Otter was a competitor for food. The various Vermin Acts of the Tudor period justified acting against that competition, and the acts' bounty payments gave an added incentive to do so.

Records of the bounty payments for the killing of the species listed in the Vermin Acts have survived to the present day. Although these records aren't complete, and those that survive haven't been completely analysed, they still provide a valuable insight into what was happening to many wildlife species in many areas across Britain.

Bounty payments for the killing of Otters can be found in records from across the country; they frequently occur. But despite this frequency, and no doubt the local pride of those claiming the bounty, it is unlikely that the Vermin Acts' bounty payment scheme had any significant impact on the British Otter population as a whole. It would have led to localised extirpation at times, but the Otters from the wider river population would soon have been able to recover the lost ground. However, during the eighteenth century this began to change, rapidly. The somewhat ad hoc bounty system for the control of species was largely supplanted by the growth of large estates that employed gamekeepers to suppress anything that was perceived to threaten the landowner's game and fish interests. For the first time, the persecution of Otters (and many other species) could be systematically and consistently organised. Rather than individual Otters being killed, entire local populations could be targeted and extinguished.

Although Otter hunting with hounds as a recreational pursuit had existed before the Vermin Acts were laid down – Henry II (monarch from 1154 to 1189) had his own Master of Otterhounds on the payroll – it wasn't until the late eighteenth century that organised Otter hunts with their own packs of hounds came into being. With the county of Devon playing host to the earliest known examples, they soon began to proliferate elsewhere.

The increase in persecution from the mid-eighteenth century onwards meant that by 1800 and the beginning of the nineteenth century, some rivers in Britain had already had their Otter populations extirpated. Worse was to come as the new century went on. Persecution of much of our wildlife, especially anything considered predatory, increased dramatically during the nineteenth century as more and more game estates established themselves. Shooting and fishing became increasingly popular pastimes for those that could afford it, and these individuals could also afford to employ people and methods to protect their quarry from anything other than their own guns, rods and hounds.

As Georgian Britain became Victorian Britain, Otter numbers dropped across the country. This was principally due to the increased levels of persecution, but also in part to changes in habitats wrought by agricultural improvements, especially the extensive drainage of wetlands such as the Fens, and riverside vegetation clearance in many other areas. As the Industrial Revolution kicked in, increasing levels of pollution in our rivers also added to the pressure being placed on the British Otter population, especially in urban areas and the new industrial heartlands of northern England. As the Victorian era drew to a close, even the Otter hunts noted the reduction in their quarry's population.

But, despite the constant persecution pressure and the increase in pollution, Otters were still well distributed across the river systems of Britain, although at a far lower density then they would naturally have been. It is likely that in some areas their presence was actually tolerated by landowners so that they would be able to have a day's pleasure with the local Otter hunt. As the twentieth century began, there were 23 packs of Otterhounds in operation across the country, including the Cheriton Otter Hunt that Henry Williamson followed and on which he based the many hunting scenes in his book. It has been estimated that at the beginning of the twentieth century, Otter hunts across the country killed, on average, a total of 430 Otters per year. Now on the face of it that number may not seem that high, but taken in the context of the time it would have been a significant number in terms of the overall British Otter population.

There was a brief respite in the killing during the First World War, when we humans turned our guns on ourselves, but the levels of persecution soon returned to their pre-war level. Following the Armistice, Otter hunting with hounds was quickly re-established and was soon flourishing, with the 1920s and 1930s generally regarded as being the heyday of Otter hunting in Britain. It was during this heyday that *Tarka the Otter* was published.

An Otter hunt was a long, drawn-out way to kill an Otter; there are reports in the hunting press of the time celebrating the pursuit of single Otters lasting for up to seven hours. Otter hunting in this way wasn't about doing a service to those who had lost fish to an Otter, or feared losing fish to it – it was about an enjoyable day out for the human participants. The day would start early, with the hunt staff checking the known holts along the stretch of river to be hunted. Because these holts were habitually used by generations of Otters, the chances were that if an Otter was present on that stretch of the river, it would be quickly located in one of them.

Once located, the hunt staff would ensure the animal remained in the holt until the hounds and the hunt followers were in a position to pursue the unfortunate creature. Using terriers and if necessary, digging equipment, they would force the Otter from the holt. They would then allow it a period of grace, a head start if you like, before the pursuit began. Armed with iron-shod wooden poles, the hunt followers would attempt to form barriers to keep the Otter from escaping the confines of the watercourse, or to turn it back towards the hounds. The chase would continue until either the Otter managed to elude its pursuers or it was torn apart alive by the hounds.

It seems strange to us now to consider that Otter hunting was a popular pastime, but it was especially so a century ago – although not everybody supported it and there were protestors who would occasionally appear at a hunt to attempt to disrupt it. But most of the time, these hunts were either supported or generally ignored by the wider populace. Back then, a hundred years ago, where Otterhounds operated, Otters were found with relative ease. Otters were still regarded as vermin, as a pest species that needed to be controlled, but their presence in a river system was more likely to be tolerated if there was a local Otter hunt. Where there were no hunts, the Otters were likely to be routinely persecuted throughout the year.

Modern thoughts and past reminders

I of course do not know how an Otter would have been viewed by the residents of the first house the River Taw flows past a hundred years ago – but I suspect, given the general view of them at the time, that it wouldn't have been favourable, maybe indifferent at best. Most likely though is that the residents of the house didn't even know they had occasional Otter neighbours at the bottom of their garden; they are after all hard to detect even when the population is thriving. It is something I can only guess at. But today, I can find out how Otters are viewed by the owners of the first house on the river, because I can go there and ask them.

It is the meteorological first day of spring as I head up the country lane to the village of Belstone on the northern edge of the moor, but someone clearly forgot to tell the weather; as I ascend, so the temperature descends until finally settling at just one degree Celsius. After the village, the lane I'm following narrows before becoming a stone-surfaced track on the edge of the moor itself. The rain of lower down has now been replaced by wet snow, the sides of the track I'm driving on slowly turning white. The huge mass of Cosdon Hill in the background is already peppered with a partial covering of snow, but the even higher ground of Steeperton and towards Taw Head isn't visible, for both are lost in brooding black clouds. As I get out of the car, the wind quickly reminds me of the inadequacy of my four layers of clothing when compared to the fur of an Otter. It is bitterly cold.

But the welcome I receive from Peter Hawes is very warm, and I soon find myself inside and enjoying a cup of hot coffee. The first property on the river is today known as Bernard's Acre, but a hundred years ago the then newly built property went by the name of Ladybrook, and was the holiday home for a family from Exeter. It was built in 1921 and even had a swimming pool. Peter tells the tale of how the lady of the family used to travel by rail from the city of Exeter to the town of Okehampton, before travelling on to Belstone by carriage to spend the summer in this beautiful spot alongside the Taw.

Today the property is run as a residential centre and it was named Bernard's Acre in memory of Bernard Partridge, a church youth leader who sadly died before he was able to realise his dream of opening the centre on the moor. Peter Hawes is one of the trustees overseeing the centre, and he tells me about the groups that visit and how the centre's location is its main selling point. We chat about how important it is for people to have access to nature and how the centre's position is ideal for people to experience the moor and its wildlife. I ask Peter how he feels about the fact that Otters occasionally use the river that forms the property's eastern boundary. He thinks it's brilliant and is rather pleased that they pass through, past the property that he helps run; he says that Otters being here is an indicator to him that all is well with nature. He is also sure that the guests that use the residential centre will be thrilled at the thought of Otters passing unseen at the bottom of the garden.

It is no surprise to find that Peter has a positive attitude when it comes to Otters – they are nowadays popular animals, after all. But a little further downstream, a relic from the past gives an all too clear reminder of how many people used to view them a hundred years ago.

The beautifully named Sticklepath is the next village down the Taw from Belstone. There are several businesses here, one of which is Dartmoor Auctions; every month, in their own eclectic way, this auction house holds a monthly sale of several hundred lots. Almost anything can come up for sale here, but, by one of those strange coincidences that seem to pepper our lives, on the same day that I speak to Peter in Belstone about Otters, I find myself also speaking to Tim Penrose of the auction house about them too. Or rather, about a bit of them.

In amongst the more than 800 lots that are coming up for sale is a rather grim reminder of the contempt in which we have held Otters in the past. It is, of all things, a brooch. A brooch made of an Otter's paw. Today, this stuffed appendage looks gruesome and hideous, but to the person who first owned it, this was evidently something they took great pride in, for they mounted the dismembered paw in solid silver. Not only that, but they recorded the paw's previous owner in engraved detail.

The paw belonged to a dog (male) Otter weighing 21 lbs, presumably including the paw. This male Otter was posthumously named Snapper by those that killed it – Williamson's favourite Otter hunt, the Cheriton Otterhounds, on 14 June 1906. I have no doubt that back then this object was worn openly and enthusiastically and attracted admiring looks, but today I find myself looking at it with sadness and revulsion. I am here to see it for myself, having noticed it in the auction catalogue. Dartmoor Auction's Tim used to be an undertaker; he is used to macabre things, but even he is a bit lost when it comes to this object consigned to his auction house from a house clearance. We're probably both thinking the same thought. Let a silver scrap dealer buy it; that way the silver will be melted down and recycled, whilst the sad trophy of a paw will be lost forever.

As I talk about Otters and auctions with Tim, he tells me that he once arranged a house clearance for the family of the writer Gavin Maxwell, coming across several first editions by him in the process. Life is strange sometimes; just an hour or so previously I had been talking to someone who was delighted to have Otters passing by the property he looks after. I followed that by viewing a grotesque reminder of how people used to treat Otters in the past, and then I was discussing a man whose books helped change the public perception of them. All that in an hour and a few miles of Tarka's Taw.

Gavin Maxwell had a relatively short and very much incident-packed life during which he wrote *Ring of Bright Water*, an autobiographical tale based mainly on his time living in a remote coastal property in western Scotland.[2] Although the book is often thought of as being about Otters, they don't actually get a mention until over a third of the way into the book; but from that moment on they are indeed the central theme. The book seems dated nowadays, sometimes even more so than *Tarka*, despite being written over thirty years after Williamson's book. It was written in the late 1950s, a time when people could apparently keep whatever pets they liked – at one point in the book Maxwell recounts the tale of an impulse purchase of a Ring-tailed Lemur that he happened to spot in Harrods one day, a purchase that nearly cost him his life.

The ottery tale in Maxwell's book is about three wild-born otters that were then taken from the wild and either sold directly to him or to others who then passed them on to him. These are not British Otters though, and not even of the same species; they are otters taken from Iraq and West Africa during the time of British colonialism. A time when it was also apparently very easy to just import a foreign wild animal into Britain without any problems or paperwork.

Unlike Williamson's book, *Ring of Bright Water* focuses on the relationship between Maxwell and the various otters he looks after. It is a book about the joy they brought to him, joy recorded in beautiful descriptive passages, rather than a book about how they are hunted – there are no ripping teeth of Otterhounds in this tale. But that being said, two of the three otters in the book die due to the actions of humans. The first, whom Maxwell named Chahala, died after (probably) eating fish that had been deliberately poisoned by fishers so that they would be easier to catch. The second, and the first to actually make it to Maxwell's house in Scotland, Mijbil, died due to the same reason as many other otters in Williamson's book: because it was an otter. A man known by the sobriquet of Big Angus was driving his lorry one day when he spotted Mijbil running alongside the road. He stopped his truck, got out a large and heavy pickaxe and bludgeoned the creature to death, simply because it was an otter. Thirty-plus years may separate the two books, but it is evident that attitudes to otters hadn't changed that much in those three decades.

Ethically there are many issues with *Ring of Bright Water*, not least the fact that Maxwell was prepared to take animals from the wild and keep them as pets. At one point, bereft at the loss of Mijbil, Maxwell describes how he searched the known holts of local Scottish Otters along the coast near where he lived in the unfulfilled hope of finding a cub he could purloin and call his own We have to remember, however, the period in which it was written when we read it today, and we also have to consider its impact. The book is often quoted as being an inspiration for a generation of naturalists and there is no doubt that it is a natural history classic, selling over two million copies since it was first published.

Even though its otters are not the same species as our own, the book had an impact on how people perceived otters as wild creatures. People like Big Angus may still have been common when the book was being written, but once published, the book's eloquence in describing the joy that the otters gave Maxwell and others helped to change how people thought about these wonderful wild creatures. Maxwell was a great writer; his prose created pictures for readers to imagine. Describing watching one of the otters swimming underwater, he wrote, 'he was boneless, mercurial, sinuous, wonderful'. It was passages like this that not only changed how people thought about otters, but made them want to venture out and see them for themselves. Ironically, as they headed out to do so, they found that in many parts of the country they were already too late.

Today, public perception of the Otter is very different to what it was a hundred years ago in the 1920s, when Williamson was writing his most famous book. You are very unlikely to see someone wearing a severed Otter's paw with pride today, but you will see people wearing images of Otters, be they somewhat garish Disneyesque T-shirts or more lifelike subtle designs. Metal pin badge images of Otters can be frequently found in a wide range of shops, surely a vast improvement on the silver-mounted monstrosity offered at the auction in Sticklepath.

Otters are a present-day marketing tool. A visit to the garden centre that lies on the west side of one of the Taw's tributaries, not far from where I am sat now, sees me immersed in their image – they adorn a plethora of consumer products. I can buy a real photographic card showing an adult and a cub looking indescribably cute; there are tea towels with swimming Otter designs, a travel mug with an Otter chasing a fish and a range of pottery items painted with numerous small Otter images. A trip to the village shop nearby reveals another product, Otter Ale from Otter Brewery, a brewery that gets its name from the nearby East Devon river rather than the animal, but even so the label of the beer bottle is dominated by an outline design of an Otter. Further afield in Bradford there is a company making drysuits that have named themselves after the Otter, an animal whose own clothing of fur keeps it dry when it is in the water – an appropriate choice of name. Not all are so appropriate: there are Otter candles, Otter knives and even Otter phone cases, products that have no relation to the animal they are named after. The name of Otter has become sellable. Otters have become a brand, something that surely would have been inconceivable in 1927.

But it is not just the Otter itself; Williamson's character Tarka has also become a brand. Today I can follow the Tarka Trail, a combination of public rights of way that allows me to walk, and in some places cycle, through the Two Rivers Country of the book. If I am not feeling quite so energetic I can sit back and instead let the train take the strain, hopping on to the Tarka Line rail service, today run by Great Western Railways,[3] which offers frequent trains between Exeter and Barnstaple, stopping at eleven stations in between and criss-crossing the River Taw as it goes. Both the trail and the railway line promote the area to visitors, and both are eagerly bought into by local businesses keen to benefit from the association.

In the 2020s it is clear that the image of an Otter has very positive associations; it is an image, a brand, with which companies and businesses want to align their products or services. Otters sell. The 1920s image of them as a quarry to hunt for our enjoyment has long gone. The Otter is no longer generally portrayed as vermin, but as we will see later, there are still some that view them in this misleading light.

The relationship between humans and Otters has shifted. Today, from the human point of view at least, it is very largely a positive one. The majority of people like Otters, and a big part of that affection for the species comes from the successes of both Williamson's *Tarka the Otter* and Maxwell's *Ring of Bright Water*.

CHAPTER 3

Poisoned Arteries

The harsh, alien, rattling shout of a cock pheasant cracks the stillness of the frozen air, but nothing responds to its sonic intrusion. The sharp chill of the early morning seemingly subdues all but the instinct to survive in the other inhabitants of the broad valley.

The out-of-place pheasant, inserted into this landscape without consent, utters forth once more, this time emphasising its call with a vigorous and noisy shake of its wings. A loose straggly group of Rooks, accompanied by their smaller relatives the Jackdaws, move purposefully across the sky high above me. Below, as I look out at the landscape unfurled there, a gossamer-like mist is clinging to the valley floor, the course of the river delineated by its shroud of vapour. As the Taw leaves the moor and passes through Mid Devon, it does so through a scene of bucolic beauty, a classic timeless rolling landscape of hedge-lined pasture forming irregular patterns.

There may be irregularity in the layout of those hedge boundaries – their delineations will have hardly changed in the last hundred years – but there is no longer any irregularity in the hedges themselves. These once hand-laid boundaries and barriers are no longer cut with the skill and care of the bill hook but are instead geometricised, standardised and brutalised by mechanical tractor-borne flails. The art of hedge laying is now virtually extinct. From a distance, these hedges may look as they did in Tarka's day, but on closer inspection they reveal themselves to be nothing more than broken torn vestiges of what was once a rich and important wildlife habitat providing cover and corridors to an abundance of species. The mechanical assault of repeated flailing has now left very visual gaps in the hedges' structure, symbolising the sad voids in their diminished communities of flora and fauna.

Farming has changed dramatically in the last hundred years. Of course it has, everything has changed; our society and the way we lead our lives is very different now to what it was a century ago. Farming is no different. So much has altered in the last century; when Williamson wrote of Otters on the Taw the idea that humans would walk on the moon wasn't just fanciful, it was completely ridiculous. But now, a hundred years later, humans walking on the moon are a mere footnote in history for many of my generation, born after the event itself.

Despite a hundred years of such profound transformation, we still today look upon landscapes like the one in front of me and think of them as being unchanged, somehow frozen in time like the beads of moisture on the frosted grasses of the field in which I stand. But all has changed, farming has changed, and changes bring consequences.

As Williamson penned his book, so farm life in Britain was evolving, albeit slowly. The First World War had seen horses, the essential living engine of British farming, requisitioned for the war effort along with much of the human farm workforce. Like those humans, many of the horses were never to return. At the same time, early mechanical tractors were beginning to appear; when the Americans joined the war in 1917 they also brought with them their tractors. When the war ended in 1918, and the Americans travelled back across the Atlantic, some of those tractors stayed behind. The mechanisation of farming had begun.

But it was a slow beginning, especially in areas like Mid Devon, where the landscape of rolling hills would prove a difficult challenge for the fledgling tractors. For the first decade after Tarka's journey down the Taw, agriculture in Devon changed very little, the relationship between farming and wildlife continuing to bump along as it had done for many years. But at the end of the 1930s the Second World War brought more upheavals to our farming landscapes. The need to increase food production became a vital national drive, and as a result, agriculture began to intensify.

Everybody was expected to do their bit for King and Country; the number of allotments in Britain doubled during the first years of the war as people followed the propaganda poster's famous slogan, 'Dig for Victory'. People everywhere strived to grow their own food to supplement that produced by farming. The farms themselves were expected to use every scrap of land they had – areas that previously had remained fallow, or as buffers, were ploughed up and planted, whilst wet areas that had always formed part of the wider riverine habitat were drained. Competition for resources from nature became something that needed to be eradicated.

With this sudden, widespread and extremely thorough intensification of agriculture, it was inevitable that the wildlife that shared the farming landscapes of Britain would be impacted. Just how great an impact this was, though, is not known. To put it simply, there were no resources at a national or local level being utilised to monitor wildlife numbers, and so there are no comprehensive records for us to check. But to give an idea of how seriously the threat of natural competition for food was taken by the authorities, let us momentarily leave the subject of Otters and take a look at a tree.

The Spindle is a common, if somewhat overlooked, component of our native tree flora. It is a small and beautiful tree that provides a wealth of late summer colour with its vivid fruits, followed by another splash of colour when its leaves turn red in the autumn. As a native tree it is no surprise that it acts as a cog in many of the ecosystems of our country, providing nectar for insects and berries for birds, as well as inadvertently playing host to a plethora of creatures that feed off it. One of those creatures is a small sap-sucking bug commonly known as Blackfly.

But Blackfly has another name, the Black Bean Aphid, and as that name suggests it likes to feed on beans, be they runner or broad. It also likes to feed on sugar beet, spinach, celery, potato, carrots, tomato… If you have a vegetable bed or an allotment, you will have probably come across these small insects; they can be a bit of an irritation to the home horticulturalist. But if you are a government trying to increase food production during wartime, they can be viewed as more than an irritant, and can even be seen as a threat to national security.

Like many insects, the Blackfly has a relatively complex life cycle, utilising different habitats for different generations at different times of the year. One of those generations lays its eggs on the woody stems of trees and shrubs, and the tree most favoured for this in large parts of Britain is the Spindle.

The British government declared the Black Bean Aphid to be a serious threat to our food production, and therefore to the nation's security. As a result, they identified the Spindle as being the chink in the armour of this marauding, food-consuming invertebrate and decided that in order to combat the aphid, they would need to fight the tree. The British government, despite the significant pressures on it due to the imminent threat of Nazi invasion, actually diverted manpower and money to draw up plans to eradicate the Spindle from Britain. It never happened;[1] I personally doubt that it could have been achieved, but to even attempt to do so would have cost so much in terms of money and labour that it beggars belief it was even considered. The fact that it was perfectly illustrates the mindset at the time towards any wildlife assumed to be in any way detrimental to efforts to increase food production.

That mindset would have affected everything in the countryside, including Otters. It is difficult to fully assess, due to the lack of any records, but the increased intensification of agriculture close to rivers as part of the push for more food production would have inevitably degraded Otter habitats. The fact that Otters had historically been seen as competition for food would no doubt have led to direct persecution of them too – especially as many of the Otter hunts stopped during the war years, removing the 'keep the Otters for the hounds' argument for not killing them. The fog of war clouds what actually happened to Otters and much of our other wildlife during the 1940s, but one thing is certain: it was nothing compared to what was to come in the 1950s.

After the Second World War, agriculture changed dramatically; mechanisation became the norm, and the days of horse-drawn ploughs and teams of manual workers were rapidly coming to an end. By the 1950s the number of horses utilised in agriculture had declined by 90% compared to the 1920s. As they went, so the tractors came, and in that same time frame the number of tractors on British farms increased by a staggering 2,000%.[2]

But it wasn't the tractors that were the problem. Instead of making plans to remove trees to keep invertebrate 'pests' down, we had developed a new way of dealing directly with the pest itself. Pesticides.

And boy were they good – they were absolutely lethal to the pests they targeted. Unfortunately, they were also absolutely lethal to everything else they came into contact with. Dieldrin tends to be the name everyone thinks of, but that particular pesticide was just one of a new wave of related compounds known as organochlorines that began to be introduced into agriculture in the second half of the 1940s. They were the go-to solution for a whole host of problematical pests, from weevils in your root crops to larvae in your cereals and ticks on your sheep. The new pesticides promised a new dawn. Their two main uses were as a seed dressing on cereal crops like wheat, and in sheep dip; but we also used them to mothproof fabrics, including the carpets on the floors of our homes and even the blankets we used on our beds.

These fabricated chemicals came with a set of statements on not only their effectiveness, but also their safety. The first part was correct, but the second was as much a fabrication as the chemicals themselves. Dieldrin and the like are indeed effective, but their toxicity isn't limited to invertebrates. They are also lethal to vertebrates such as birds and mammals (including us). They are also extremely persistent in the environment; these fabricated chemicals don't break down into safer compounds but stay just as lethal as ever, forever. The result of their widespread usage across our country was complete decimation of large swathes of our wildlife, including Otters.

As the Second World War ended in 1945, DDT and a compound with the less memorable name of benzene hexachloride were introduced. These were used in orchards to help safeguard and maximise fruit production, defending the crop from a variety of invertebrate species. But their use did not stop there – they very quickly became more widespread in application, soon being utilised in a wide range of crop protection situations. Before the end of the 1940s, just a few years after their introduction, their usage had diversified from plant crops to animal ones as well. As the decade ended, they were being used in sheep dips across large parts of the country. DDT in particular was a popular choice for this chemical cocktail bath, used to control ticks and blowfly larvae. This chemical onslaught against anything perceived as a pest only intensified; new pesticides were being developed and marketed all the time, a chemical agricultural arms race was in full swing, and by the mid-1950s a new and 'improved' group known as cyclodienes were released onto the market. These products included the now infamous names of dieldrin, aldrin and heptachlor.

The development of these substances was widely celebrated; proud full-page advertisements in everyday magazines, not just agricultural ones, were taken out, proclaiming how human ingenuity, when it came to dealing with pests, knew no bounds. These new chemicals were widely used, especially as a seed dressing in a process that involves coating the seeds with the chemicals before they are then sown in the fields. As with DDT before it, dieldrin was also used in sheep dips, gradually replacing the former as the years rolled on. As the 1960s dawned, as Gavin Maxwell's *Ring of Bright Water* was published, as its readers thought about seeing Otters for themselves, all aspects of British agriculture were awash with these new pesticides – as was, often quite literally, the countryside. It wasn't just in Britain that this was happening; the organochloride chemicals rapidly became established in ecosystems far and wide right across the globe. Their pernicious persistence meant that DDT residues weren't just present in farmland areas where they had been directly applied, but also in areas where farming had never existed, including even in Antarctica.

It did not take long for their lethally toxic effects to be seen. As early on as the late 1950s, the use of cyclodienes as a seed dressing was causing very noticeable mass deaths among species of seed-eating birds. Huge numbers of carcasses were being found in fields recently sown with the chemically laced crops; it should have been immediately obvious to all that these new wonder chemicals were not in fact safe to use and that they should be immediately withdrawn. Sadly, and devastatingly for much of our wildlife, vested interests and that all-too-human trait of being unwilling to admit that we have got it badly wrong meant that these chemicals continued to be used for far too long, even after it was obvious that they were having devastating effects on wildlife.

It took until 1962, the year in which *Silent Spring* by Rachel Carson was published, for dieldrin, aldrin and heptachlor to be banned for use as a seed dressing – but even then, it was a voluntary ban rather than a legal one. Although most of the products containing the three substances were withdrawn from sale and manufacture, it was not until 1981 that the use of dieldrin in Britain actually became illegal. Heptachlor followed in 1984, but it took until 1991 for the use of aldrin to become illegal in this country.

Biomagnification is a big word that has big consequences, especially for predatory species such as Otters. It is a word used to describe the process in which increasing concentrations of a toxin, such as a pesticide, accumulate in the tissues of species at successively higher levels of the food chain. The species at the top of the food chain, species like the Otter, are those that end up accumulating dangerously high levels of the toxin. The low levels found in the species at the beginning of the food chain may not have any adverse effect on that species, but as they get eaten so those low levels increase in the species that is the next link in the chain. On it goes, right through the food chain, toxin levels accumulating at each stage, until it reaches the apex predator, the species which gets the highest dose. This accumulation of toxin in the apex predator can cause an immediate death, or a long lingering one; or, if not quite at lethal levels, it can seriously impair the animal's bodily functions, especially its breeding ability, which in turn leads to population decline over time.

Because Otters are elusive creatures, the impact on them caused by the widespread use of cyclodiene compounds in agriculture was not immediately obvious. By the time Gavin Maxwell's *Ring of Bright Water* was climbing up the bestseller list and changing people's perceptions of Otters, the wild Otters of Britain were being hit hard and disappearing at speed. Their rapid demise across the country simply wasn't being noticed; when a species has a reputation for not being seen, it makes it very difficult to know when there are none left to be seen.

But there was one iconic predatory species whose rapid decline, due to the use of these pesticides, was noticed and recorded, and that was the Peregrine Falcon. These majestic raptors are obligate bird feeders, the ultimate avian predator in both senses of the words; they hunt a wide variety of bird species, and many of those species are seed-eating ones such as the Woodpigeon. The Peregrine has long held the fascination of us humans, and as a result their presence in areas has long been noted and recorded, though this hasn't necessarily led to them being protected from our actions.

The Woodpigeon is not the only pigeon these raptorous birds predate; they will also hunt domestic pigeons,[3] including those that were used to send messages back to Britain from occupied Europe during the Second World War.[4] This put the Peregrine at odds with our war effort, since pigeons played a vital part. The successful delivery of the messages they carried was so important that the falcon was deemed a real threat; whilst many species, including the Otter, were routinely persecuted during the war, the Peregrine was one that was specifically targeted by government legislation. In 1940, less than a year into the war, the Destruction of Peregrine Falcons Order came into effect. It ran from July 1940 through to February 1946,[5] and was an official licence to kill these birds and destroy their nests. It was a licence that was readily used. Although an accurate figure of the number of birds killed is difficult to ascertain as the record keeping was slack, it has been estimated by ornithologists that around 600 adult birds were killed. In addition to that number, there were of course countless young and eggs that were destroyed in the nests of the species at the same time. This somewhat unsurprisingly had an impact on the population of the Peregrine in Britain, and by the time the order was finally lifted in 1946 it is believed that their population had fallen to around 87% of what it was in 1939. Only 85% of known breeding territories in the 1930s were still occupied by the birds after the war.

A 13% drop sounds severe, but it was a mere drop in the ocean compared to what was to come. Once the government's destruction order was lifted and the permitted killing stopped, the raptor's population began to slowly recover, as would be expected, but what was not anticipated was the sudden reduced breeding success that started to be noticed in 1956. Observers who had been witnessing the slow recovery of the species since 1946 suddenly discovered that traditional territories used year in year out had become unoccupied, or, if they were occupied, the adult birds present were failing to rear young or in some cases not even attempting to breed at all. This reduced breeding success was first noticed in the south-west of England, but it was soon noted over a wider area, and by 1961 it had been recorded right across the British Isles, from Land's End to the Highlands of Scotland.

In 1962 the rate of occupied Peregrine territories in Britain as a whole was less than 44% of what it had been in the 1930s – a massive 56% loss in occupied breeding territories. But in southern England, where the pesticide use was concentrated, it was far, far worse. A mere year later, in 1963, there were only three Peregrine breeding territories recorded as being occupied in the whole of southern England, below an imaginary line between the Mersey and the Humber. Apart from these three territories, two of which were in the Taw's county of Devon, the Peregrine had been wiped out from this vast area of England. The licensed killings to protect our war effort seem almost paltry when compared to the pesticide-induced massacre of the late 1950s. It has been calculated by ornithologists that if nothing had been done, if the voluntary ban on the pesticides had not occurred in 1962, a ban which led to a dramatic reduction in their usage, then the Peregrine Falcon would have likely become extinct as a breeding bird in Britain by 1967. The use of this new wave of 'improved' pesticides was nothing other than an utter disaster for much of Britain's wildlife.

The decline of the Otter was no less severe than that of the Peregrine. Just as the pesticides found their way into the avian food chain, so they also found their way into the aquatic one. Any persistent chemical applied to land – be that by coating seeds, spraying crops or emptying sheep dips – at the end of the day will, inevitably, make its way into our watercourses and into the food chains within. Otters love to feed upon Eels and those Eels love to feed on a whole host of invertebrates, from worms to insect larvae, exactly the sort of species that these new agricultural chemicals were designed to target. The Eel is a long-lived creature that can spend a dozen or so years maturing in our rivers and ponds before they embark on their famously epic journey back to their breeding grounds in the Sargasso Sea in the North Atlantic. Those years are primarily spent feeding, and if the food they are eating carries with it a toxicity, then that toxicity will become ever more concentrated in the Eel as it continues to feed, day after day, month after month, year after year. Eels have relatively high levels of body fat; they are very calorific to eat, which is no doubt one of the reasons why Otters like to feed on them so much. High body fat, however, also enables the Eels to store contaminants in their bodies without it drastically affecting them. Contaminants like persistent pesticides. When an Eel is eaten by an Otter, the Otter is also consuming the fat-stored contaminants it contains, and these contaminants increase in concentration in the Otter's body each time an Eel is eaten.

But unlike the fatty Eel, the Otter doesn't have much body fat to store any of those contaminants in. As we have seen, an Otter actually has very low body fat, around just 3% of its body mass compared to up to 30% in an Eel. This means that unlike the fish, the Otter has no buffer against the toxicity of those contaminants; instead they are extremely vulnerable to them. But this toxicity in their diet did not immediately prove lethal. It did not, typically, kill them outright – people weren't finding the bodies of Otters washed up on riverbanks, bodies that might have indicated that something was going badly wrong. Instead of outright dying, Otters were dying out. The toxic load in their bodies badly affected their reproductive organs, and they were no longer able to successfully breed – they were no longer producing new generations. Otters are relatively short-lived creatures; they can live up to ten years, but the vast majority of them never get near that lifespan. As the Otters in our rivers were dying in the 1950s and 1960s there were no new young Otters being born to replace them, and so they were becoming functionally extinct.[6]

The hunt for answers

Whilst this insidious decline of the Otter went largely unnoticed by society, it wasn't completely unobserved. One part of society was indeed aware and increasingly concerned. Otter hunting with hounds had stopped during the Second World War, but it didn't take very long to restart after the conflict had ended; however, for a variety of reasons, including an increase in pressure from anti-Otter-hunt campaigners, the number of active Otterhound packs had diminished somewhat from the 23 of the pre-war heyday to 12 in 1951. For a while, for the hunts that remained, things carried on much as they had before. Otters were to be found where Otters had always been found, and were therefore hunted – nothing had changed. But then, from the mid-1950s onwards, it did change.

The hunts rapidly noticed that Otters were no longer being found in their usual haunts; instead of finding their quarry in the traditionally used holts along a river system, they were drawing blanks. At first this was thought to be indicative of local declines, something that was only happening on their 'patch', and indeed some hunts even took the not inconsiderable step of importing Otters from abroad to restock their rivers. But it soon became evident to the hunts that this was not just isolated local declines – something else was clearly at play. By the beginning of the 1960s Otter hunts in all parts of the country were struggling to find any Otter to hunt.

In 1962, in the now defunct *Gamekeeper and Countryside* magazine, an article was published entitled 'Where Are the Otters?'. Written by an Otter hunter, it told the tale of how hunts across the country had hardly found any Otters over the last few years; something, the article stated, was clearly wrong. The article's author then went on to speculate that habitat changes caused by our overenthusiastic clearance of rivers to increase drainage – especially the removal of riverside trees, willow clumps and reed beds – combined with an increase in water extraction, which in turn had led to lower water levels, had caused the decline in their quarry. But this was not the only theory proposed by the Otter hunters. Others among them suggested that it was people visiting the countryside that were to blame, that too many picnickers and bathers had simply scared the Otters away. It is a suggestion that brought a wry smile to my face when I first read it. There is definitely something ironic in the thought that a family picnic was more terrifying for the local Otters than a pack of hounds intent on killing them!

Their theories may have been wide of the mark (although we will return to the subject of work done on rivers to improve drainage later), and their motivation for questioning what had happened to the Otters may itself be questionable to people like myself. There can be no doubt, however, that the Otter hunters were genuinely concerned by the drastic decline in their quarry. Indeed, they were so concerned by it that they subsequently changed their hunting policy. Prior to the noticed decline in the numbers of Otters, the hunts would typically kill 50% of all animals they found during a hunt. During the 1960s the hunts made the decision to reduce this number down to 15% of animals found, a self-imposed policy that continued right through until hunting Otters was finally banned. They also changed their modus operandi and instead of looking for an animal that offered them 'good sport' – in other words, a prolonged chase – the hunts opted to only kill animals they thought to be injured or sick (of course, *all* Otters were sick by this stage!), or those that they had been specifically requested to hunt by a landowner who thought that a particular Otter was causing them problems.

Fifteen per cent of not very many meant that instead of killing an average of 430 Otters a year across the British Isles, as they had done previously, the hunts on average were killing a total of just 11 animals per year. These very low figures, and the increasing clamour of the concerned Otter hunting community, caused scientific and conservation groups to also become increasingly worried about the reported decline. The problem for the scientific community was that hard data was hard to come by; no one seemed to know how many Otters there were supposed to be. A report by the Mammal Society in 1969 concluded that there had indeed been a serious decline in the number of Otters, but the report was an anecdotal one based on interviews with people and river authorities, rather than the sort of robust data that scientists crave and policy-makers require.

But the Otter hunts themselves were able to provide that hard and fast data – although initially it was only partial, due to a lack of coordinated record keeping. But even though incomplete, that primary tranche of data clearly showed that there had been an average success rate of around 70 Otters killed per 100 days hunting before 1957, but by 1967 that success rate had fallen to 44 Otters killed per 100 days. This information enabled scientists to say with certainty that somewhere in the decade between those two dates, something had happened to cause the reported decline, but they weren't sure what that cause was. It seems strange to me, looking at the data now, that they weren't able to join the dots at this stage – after all, it was already well known that the new wave of pesticides introduced during the mid-1950s were having a devastating effect on wildlife such as the Peregrine Falcon. But despite this seemingly glaring potential cause, pesticides were just one of six proposed factors that the scientists suggested for causing the decline in Otter hunting success rates. The other five proposed causes were the severe winter of 1962/3 (one of the coldest ever on record), increased tourism, increased fishing, destruction of habitat and fur trapping. The waters were further muddied when, ten years later and with Otters now extinct in many parts of the country, another report suggested yet more potential causes: the spread of the American Mink, an unknown disease and road casualties.

Because the actual starting point of the decline had not been pinpointed, the link to the introduction of cyclodienes like dieldrin couldn't be proved, but thanks to Otter expert Paul Chanin it finally was. In 1977 Paul managed to ferret out a complete dataset of Otter hunts from the year 1950 through to the year 1966. It was this dataset that enabled him and fellow Otter expert Don Jefferies to pinpoint the start date with much more accuracy; using this new data they were able to show that the decline had started in 1957, immediately after the widespread introduction of dieldrin, aldrin and heptachlor. This starting date, coupled with the fact that the data showed the decline had started in the southern half of the country (the same had happened with the Peregrine) and had been rapid and almost simultaneous, meant that Paul and Don were able to discard many of the other theories that had been proposed. They concluded that the severe decline in Otter numbers was down to the introduction and widespread use of the new wave of agricultural cyclodiene pesticides. They were of course completely right.

By 1977 the Otter had been virtually wiped out from England; it had completely vanished from entire river systems across vast swathes of the country, river systems where it had been present for millennia. They did somehow hang on in a few places – Devon and Cornwall held the strongest populations (the Taw did not lose all of its Otters), whilst some also survived in the Welsh Borders and a few hung on in North Norfolk and in Northumberland, but that was it. Ottery places had been rendered otterless almost everywhere in England. Scotland and Wales held better numbers but even these were reduced and, in some areas, particularly in the south of the two countries, the Otter was now absent.

That link to the introduction of the pesticides was demonstrated in 1977 and the authorities of the day took immediate action, introducing full legal protection for the Otter in 1978 which effectively banned Otter hunting.[7] Now, whilst I am not defending the hunting of Otters with hounds (to be perfectly clear, it is something that I find completely abhorrent), I do find it frustrating and somewhat ironic that this barbaric practice was only banned because of the virtual wiping out of Britain's Otters due to agricultural chemicals; and that those chemicals, the actual reason the Otters had been virtually wiped out in the first place, weren't banned until much later – in 1991 in the case of aldrin. Giving the Otter full legal protection was of course a great thing to do, and it certainly made for a pleasing headline, but the real priority should have been the banning of the deadly toxic compounds that caused their catastrophic decline, chemicals we were so freely using in our food production.

Whilst by 1977 the use of the organochlorine family of pesticides such as dieldrin and aldrin had been phased out (but not actually banned, don't forget), it didn't mean that things were looking up for the Otter (or other wildlife). Stopping the use of organochlorines was indeed good news for Otters, but as is so often the case, there was very much a cloud within that silver lining. If the agricultural industry couldn't use organochlorine compounds on their crops, they needed to find a viable replacement, and so agriculture started using organophosphate compounds instead, compounds that were almost as bad as the ones they replaced. Unsurprisingly it wasn't long before these too found their way into our river systems, no doubt helped by the fact that sheep dipping with these chemicals was actually made compulsory until 1992, the year in which, finally, organo-chemicals, whether phosphates or chlorates, were phased out for good.

But such is our desire to liberally pepper our food-producing agricultural sector with synthetic chemicals, the organophosphates were simply replaced with a new creation, pyrethroids, a group of compounds that, unlike their predecessors, did not accrue up through the food chain. Instead of doing so, this new group of fabricated compounds simply broke the food chain completely, wiping out entire groups of invertebrates. Once these new chemicals invariably found their way into our river systems, they took out the invertebrates present in them, invertebrates that were of course the food for many species of fish, fish that in turn were the primary food for Otters. Food chains are so called for a very good reason: they are composed of links, and if you break the links of a chain then it fails to work, it falls apart. Thankfully, in 2006, 50 years after the likes of dieldrin were introduced into agriculture, the pyrethroids were also banned. Fifty years of chemical abuse, fifty years of chemically polluted waterways, fifty years of poisoned arteries.

It is of course tempting to think that we have perhaps learnt our lesson.

But before we look at whether we have, it is worth examining how species are able to regain ground once the pressures that are affecting them have been relieved. It is also worth pointing out that, although disastrous for many species, the toxic poisoning of our waterways in the second half of the 1950s was not a new phenomenon. We had been doing it for centuries, and we had especially been doing it in our urban centres of industry.

Coming back

Nature has an innate ability to bounce back if allowed to do so; as a Greek philosopher supposedly once said, nature abhors a vacuum. The reduction in use and eventual withdrawal of the various organo compounds in the agricultural sector meant that many species that had been seriously threatened with extinction across large parts of the country were now able to fill those vacuums; they were able to recover, and in some cases even expand, their ranges. The Peregrine Falcon, the magnificent raptor we looked at earlier, is a good example of how a species can come back from the very brink, and whilst its recovery since the early 1960s was no doubt also enabled by a reduction in persecution (although, sadly, this is something that still occurs today, despite the bird being legally protected), it was the removal of the chemicals from our countryside that allowed the bird to rather dramatically reverse its severe decline.

Today, the Peregrine Falcon is not a rare bird in Britain and for many people it is hard to imagine that it once was, even within some of our own lifetimes. Their recovery does not, nor should it, detract from the excitement of watching one in action, however; they are thrilling birds to see and today many people in the UK could do so, if only they opened their eyes to the opportunity. The nadir of just three pairs in the whole of southern England in 1963 seems ridiculous to us now; it sounds unbelievable. Especially when we look at the numbers of Peregrines today.

In 1963, two of those three breeding pairs were found in my (and Tarka's) home county of Devon – two thirds of the entire southern England population. But in the most recent British Trust for Ornithology (BTO) survey of Peregrines, a total of 93 (yes, 93) pairs of this dashing raptor were recorded in Devon,[8] a truly remarkable comeback. Without the chemicals killing them and reducing their breeding ability, the birds have returned in number and, in doing so, they have also mirrored something many humans have also done since the 1960s. They have moved from the countryside into the cities.

Urban Peregrines are now very much a fixture of our towns and cities right across the country, from Devon's county capital of Exeter to the national capitals of London, Cardiff, Edinburgh and beyond. These are no longer birds to only be found in wild inaccessible countryside; these are birds that can be found on the doorsteps of millions of people. If those people looked up a bit more, they would see Peregrines. Our tall urban buildings make fantastic nest sites, acting as purpose-built cliff faces ready to be used. But the Peregrine isn't the only species that was facing extinction just a few decades ago and yet can now be found living in our towns and cities. Otters too have become urbanites.

When we think of rivers, we often picture bucolic scenes, babbling waterways flowing through verdant countryside – but these rivers also flow, inevitably, through urban areas, be they the Thames or the Taw. Whilst the town of Barnstaple is a very different place to London, the River Taw becomes an urban one when it flows through it. In the 1950s and 1960s the agricultural chemical-laden rivers of our countryside were also agricultural chemical-laden rivers of our towns and cities; the water, and all it carries with it, flowed through them regardless. Whilst these chemical pollutants originated from our farmed countryside, they would not have been the first pollutants in these urban waters. Britain, particularly its industrial heartlands, has a long history of urban rivers being polluted with contaminants arising from the built environment, and these contaminants were just as disastrous for wildlife.

I can remember as a child visiting family friends in Yorkshire in the mid-1980s; we spent a day in the city of Bradford where we visited museums and art galleries, but it was not the museum exhibits or the David Hockneys that stuck in my mind. It is a sign, or rather several signs, located alongside the open urban sections of the city's waterways that vividly replay in my mind.[9] The signs warned you to keep away from the water, not because of a risk of drowning, but because they were dangerously toxic. Bradford and many other Northern towns and cities in England were industrial centres in the Georgian, Victorian and Edwardian eras. They were full of mills, factories and poor-quality housing; any waterways within them, be they brooks, rivers or canals, were simply treated as open sewers and industrial drains. They became very dangerous places.

The Bradford Beck is the river that flows through the city, but it is mostly hidden from view nowadays, built over for much of its urban windings. It is still there, though, and occasional stretches are accessible; it was along these stretches that I saw those shocking warnings. Whilst the signs were relatively modern in the 1980s, the toxicity of the water dated from much further back. In the early 1800s the Bradford Beck had the unenviable sobriquet of the 'filthiest river in England'. The German philosopher Friedrich Engels, after visiting Bradford, described it in the 1840s as being a 'coal black, foul smelling stream'. The waters from the beck were used to top up the waters of Bradford's canal, a conduit to the larger Liverpool and Leeds canal. As a result of this topping up from the beck, and of the horrendous pollution being fed directly into the canal, this human-created waterway actually ran the risk of igniting. When stretches of water are at risk of bursting into flame, it is pretty clear that something has gone seriously wrong; canals are not meant to be flammable![10] Such was the real and continuous danger of this happening that the decision was taken to close the canal in 1866 because it was seen as too great a public health hazard. Bearing in mind the conditions of Victorian England and their very scant attitude to what we would today call health and safety, the fact that the canal was closed for this reason speaks volumes about how very badly polluted it must have been.

Otters would have once used this waterway, taking advantage of the new canal systems that were being built across the country, but they would not have survived for very long in Bradford's canal and beck waters once the Industrial Revolution poisoned them. When the agricultural chemicals that decimated Otters across the country from the 1950s onwards flowed into the beck, as they inevitably would have done, they would have had no effect at all on any Otter population, for they were so long gone by then they weren't even memories.

Sadly, Bradford Beck is just one of many watercourses in northern England that is still living with the legacy of the

Industrial Revolution's disregard for them. Whilst steps have been taken to help clean up Bradford Beck, there is still some way to go before it might support Otters once more. However, there is hope – they are back in other once heavily polluted Yorkshire rivers.

Sheffield is a major city in Yorkshire with a population of over half a million people. It has long been a centre of heavy industry, particularly coal mining and more latterly steel making, and is a city of three rivers: the Don, the Sheaf and the Rother. All three suffered badly as the city's industry developed. Everything from raw sewage through to coal slag was routinely dumped in the rivers; the dyes and washing chemicals from the wool and spinning industry drained freely into them; and as the steel industry took hold, the waters were artificially heated due to being diverted through the plants to cool the foundries' internal workings. Artificially warm waters combined with raw sewage leads to algae blooms forming in proliferation, compounding the already dire ecological situation the rivers were facing. By the mid-twentieth century the three rivers of Sheffield were among the most polluted of all European rivers. Their ecology had collapsed, and they were effectively biologically dead.[11] As with the waters of Bradford, the Otters of Sheffield would have been long gone before the organo compounds of the 1950s arrived.

But, as we have already seen with the Peregrine Falcon, nature can and does come back if the correct conditions are created. During the 1980s, British industry was in decline, particularly in the north of England – locally the large steel, coal and chemical works that caused much of the pollution closed, whilst nationally there were wholescale changes to legislation, leading to increased policing of pollution backed up by large financial penalties. This, coupled with earlier improvements in our sewage systems, gave rivers like the Don, Sheaf and Rother the time and the space to heal themselves. As conditions improved, once lost aquatic plant communities and invertebrates re-established themselves; fish species returned, either naturally or with help from the authorities; and as the fish returned so did their predators, with bird species such as Kingfisher and Grey Heron becoming regulars along the rivers once again. They were not the only returnees. It is not known when Otters actually died out in Sheffield's rivers, but it would certainly have been a long time ago. But by the beginning of this millennium, Otters were being recorded back where they belonged, in urban Sheffield's rivers and watercourses.

This has happened across many formerly badly polluted urban waterways; the rivers that flow through our cities have Otters in them again. In Devon, the busy quay area of the Exe as it flows through the heart of Exeter is now a good place to see these mammals. Late-night revellers leaving the area's nightclubs are likely to end up very close to an Otter, even if they are too oblivious to notice!

Upstream of Exeter, the River Exe flows through the town of Tiverton, its wide expanse coursing right through the centre of it. We don't normally think of town centres as being good places for watching wildlife, but as I was writing this chapter, I received a video filmed right in the heart of the town. It was recorded in the middle of the day and shows a very relaxed male Otter rolling around on its back on a flattened couch of sedge, enjoying the warmth of some winter sun. At one point it looks up towards the camera, evidently aware that people are present, but it continues to have a good roll before eventually moving out of sight.

Today, from London to Edinburgh, Exeter to Newcastle, and Tiverton to Telford, Otters are back in our urban rivers. Like the Peregrine Falcon, Otters have made a comeback right across the country, so much so that they are once again reclaiming territories long swallowed up by our expanding towns and cities.

But it is a comeback that has taken time.

CHAPTER 4

The Return

The footpath runs straight across the grassy field that lies between the small railway station and the river. There is a wetness about the grass that gives an unpleasant reminder of the rain from the day before as it seeps through my shoes. The land I am walking across slopes gently down towards the watercourse; a large pool-like section of the river is in view and stood sentry-like to one side of it is a Grey Heron, a bird evidently not as worried as I am about getting its feet wet. I do my best to appear disinterested in it, hoping that it will tolerate my presence as I approach the bridge. But even though I am not directly heading for it, it decides to move on, opening its wings and beating them downwards while using its long legs to leap upwards into the air.

I do not go to cross the bridge, though; instead, although initially distracted by the bright yellow flash of a Grey Wagtail as it flies across the water, I focus on the base of one of the bridge's supports – a large lump of uninspiring human engineering protruding from the waters of the Taw. Here is a mixture of concrete and stone that in any other context I would completely ignore, but in this riverine situation, it is the type of thing I am always drawn to. It can tell me something I wish to know. It can tell me whether there are Otters about. This stretch of the Taw is beautiful, tree-lined and full of life; insects are in the air, birds are busy either pursuing them or singing.

But not all of the river's life shows itself so easily. Otters certainly do not, but they do leave signs.

Despite the discontinuation of pesticides such as dieldrin, Otter numbers in many parts of Britain remained stubbornly low for many years, whilst other species such as the Peregrine and the Sparrowhawk (another raptor that suffered badly due to the accumulation of pesticides in the food chain) rebounded relatively quickly. There are several potential reasons for this, and we should not forget that other chemical pollutants such as PCBs are also present in our river systems and also affect the breeding ability of animals like the Otter.[1] However, leaving chemical pollutants behind, I am instead going to focus on just two of those potential reasons.

The first is, of course, down to us. Habitat destruction is a term often used by conservationists, and one that generally conjures up images of bulldozers clearing large areas of heathland for a new housing development, or of the wholescale clear felling of large swathes of Amazonian rainforest to make way for cattle ranching. But it is also applicable to our management of river systems in Britain in the 1970s and early 1980s.

In 1973 the Water Act was passed by the UK Parliament, completely reorganising water, sewage and river management in England and Wales. The Act created ten new regional water authorities, replacing the disjointed hodgepodge of local-authority-controlled systems and the local river authorities that had previously been in place. Following their creation, many of these new regional authorities sought to improve the rivers and waterways that they now effectively controlled. Of course, the use of the word 'improve' is subjective. Many of these improvement projects were carried out on a large scale, and virtually all of them were about speeding up the time it took for water to flow from the river's source to the sea.

To help achieve this aim, thousands and thousands of bankside trees were felled and to stop them growing back again their roots were also grubbed up in the process; the banks were straightened and neatened by their removal. This had the immediate effect of removing thousands and thousands of potential Otter holt sites from river systems up and down the country, and as we have already seen, an Otter territory without holt sites is not an Otter breeding territory. But this was not the only 'improvement' carried out by the newly created water authorities – they also drained neighbouring wetlands and removed large areas of reed beds and scrubby willow carr, all of which were places where Otters could find food and shelter. Coupled with large-scale river straightening in many parts of the country, these so-called improvements rendered many parts of our river systems unsustainable for Otters.

And let us not forget that this all happened at a time when Otters were at their lowest ebb, virtually wiped out in England; some of those that had clung on suddenly found themselves confronted with rivers that offered little in the way of anything for them. There was water for them to swim in, but feeding sites, lying up sites and, crucially, breeding sites had disappeared on a large scale.

Whilst we are on the subject of how river 'improvement' schemes impact Otters, we should also recall that aforementioned 1962 letter to the *Gamekeeper and Countryside* magazine, when the lack of Otters following their population crash was first recorded in print by the Otter hunting community. The person writing that letter had laid the blame for the decline of his quarry squarely at the door of such river improvements, clearly stating his opinion that the felling of riverside trees, removal of wet areas, willow carr and so on were the likely cause of the Otter's disappearance. We now know that they were wrong in thinking that the catastrophic decline in Otters was down to this, as opposed to the chemical compounds we were letting loose all over the countryside. It still should be acknowledged, however, that they were correct in thinking that these ostensible improvements were not good for Otters. We may have made a concerted effort to 'improve' our rivers in the 1970s and the 1980s, but this letter shows us that such 'improvement' was not new. It had being going on for many years and concerns about its effects on wildlife such as Otters were already being voiced, even if they were not being heard by those that should have been listening.

Quality of habitat – whether a particular habitat fulfils a particular species' needs – will obviously impact on how that species is able to spread out and recolonise areas. With Otters, though, there is another factor at play: the Otters themselves. Otters, by their very nature, live at low densities; they are territorial animals and the territories they live in are very linear. In consequence they are a lot less mobile than other species, especially birds such as the Peregrine and the Sparrowhawk, two species that were able to bounce back far quicker. By less mobile I don't mean that they are physically less able to move; I mean that they are restricted in where they can move to.

Otters move through river systems and whilst they are capable of transferring between different ones, they are not animals able to suddenly colonise a new area that is separated from the area in which they have been reared. A young Sparrowhawk or Peregrine looking to establish itself in its own territory has the ability to fly across large areas of unsuitable terrain in any direction it chooses until it finds an appropriate vacant site, but a young Otter obviously cannot do that; it is restricted in its dispersal movements by the river system in which it resides. This means that it can take a long time for Otters to recolonise river systems that they have completely vanished from – their own river system needs to become saturated territorially before they are forced to move between watersheds. When a river system's population has become suppressed due to a population crash, it is going to take several years for that recovering population to reach that river's carrying capacity. Dispersing young Otters do not need to move to a new river system if the one they were reared in already has plenty of vacant territories within it.

Otters are relatively slow breeders, producing low numbers of young, many of which don't survive to breeding age. This means that they are always going to be slow in recovering their numbers following a population crash, especially one so severe. In areas of the country where Otters still survived following the pesticide pollution of the 1950s and 1960s, they didn't do so as thriving populations. They existed as suppressed populations, at artificially low densities. There would therefore have been plenty of unoccupied territories within these river systems, and these very local territories would have been the first to be recolonised as the population started to recover. Any young Otter looking to find its own territory after the pesticide pollution had abated did not need to colonise a new river system; it did not need to go on an epic journey, crossing watersheds, but only had to move upstream or downstream to find a vacant territory to claim. When the fictional Tarka switched from the Torridge watershed to the Taw watershed, he did so only because he could not find sufficient space on the former to set up his own territory.

Most of the 48 counties in England lost their Otters following that pesticide-pollution-induced crash; they did not have residual populations from which the recovery could begin, because they had no Otters left at all. It was always going to take a long time for the surviving Otters elsewhere to recolonise these areas, but our river 'improvements' made that process even longer. Otters need breeding sites to be present on a section of river – they cannot create their own breeding holts, but can only take advantage of ones that already occur. If we remove these potential breeding sites through so-called improvements, then we effectively make that section of river sterile from an Otter's viewpoint. Otters will and do use areas of river that do not have breeding sites. They will explore them and they will fish in them, they will spend time in them, but they will not breed in them. If they are not breeding, they are not producing young, and if they are not producing young, their population will not expand at all.

It is highly likely that the wave of river improvement schemes that flowed forth following the creation of the new water authorities in the early 1970s seriously impeded the recovery of the Otter population. It is also likely that more local extinctions on river systems took place as a result of these schemes being implemented, schemes which effectively isolated an already fragmented population of Otters. The words 'isolated' and 'fragmented' never bode well for wildlife. At exactly the time that our Otter populations needed help to recover from their nadir, we just made it worse.

Thankfully, by the mid-1980s, attitudes (and without being too cynical, perhaps budgets for large-scale schemes) had changed, and river improvement schemes in their initial form were largely scaled back or even scrapped. Since the mid-1980s, Otters have indeed recovered. National Otter Surveys have been undertaken periodically in England by the Mammal Society, and these surveys have documented that recovery, with the fourth carried out in 2002 demonstrating that the number of sites with a positive Otter presence had increased by 55% since the previous survey in 1994.

When surveying for Otters, you are extremely unlikely to see the animal itself – their elusiveness means you must look for their signs instead. With Otters that means searching for spraints, their characteristic droppings that they freely and, luckily for Otter surveyors, conspicuously use to mark their territories, both on the territory margins and throughout, with sites like the bases of bridge supports offering perfect places for such a message-laden deposit.

On paper, the data from the 2002 survey, showing a 55% increase in sites with positive signs since the previous survey, seemed like a brilliant result. It was, but it also has to be put into perspective: positive Otter presence was only detected in that 2002 survey at 36% of all the sites, meaning that around two thirds of all sites surveyed in England still did not have positive signs that Otters were present. Twenty-five years on from the link between their population crash and pesticides like dieldrin finally being proved, the Otter population of England was still far lower than it should have been.

Eight years later the fifth National Otter Survey of England was carried out, once again organised and led by the Mammal Society, and it revealed that the increase in Otter-positive sites had continued. In this survey 56% of all the sites surveyed had positive Otter signs present. This was a welcome increase on the previous survey, showing that over half of the sites surveyed now had their Otters back. But we must not forget that it had taken over five decades, from the start of their population crash, for Otters to reappear in half of England's river systems, something that really demonstrates just how close we came to losing these creatures. The 2010 survey also showed that almost every county in England, 47 out of 48, was now home to Otters again; it was only the far south-eastern county of Kent that was still lacking them. But then, a year later in 2011 on the Kentish rivers of the Medway and the Eden, occupied Otter holts were found. For the first time since the mid-1950s Otters were breeding in Kent, and thus for the first time in over half a century Otters could be found in every English county. The Otter has made a slow comeback, but it is nonetheless still a remarkable one; it has returned from the brink. All is looking good once more for our Otters.

Or is it?

But before we answer that question, we are going to meander along a couple of other themes – the first being the changes visited upon the wildlife of our rivers since the fictional Tarka's day.

CHAPTER 5

Lost, Found and Returned

A low, dense, clingy mist fills the valley of the Taw as it meanders its way northwards, flanked as it travels by the main road between Exeter and Barnstaple, a road that mirrors the curves of the watercourse. The river is broad and slow at this part of its journey, a bit like the road, and although the constant tarmac bends tell you of its presence, it only fleetingly comes into view as you drive. Stopping at Umberleigh allows me to see the river again as it unhurriedly flows beneath its ornate-looking bridge. The bridge was rebuilt completely just a few years before Tarka was published; in Tarka's day there would no doubt have been Otter hunts passing through this section of river, the hunt's followers standing on the very bridge on which I find myself today. They would have been looking at hounds and huntsmen, and the river would have been abuzz with noise from the throng of hunters and watchers. Today there is just me here, peering down through the damp mist, the only sound the soft burbling of the water flowing beneath me.

I am not alone, though – a pure white form is standing some 40 metres downstream of me, in the liminal space where the shallow water brushes against the sloped bank. The mist obscures its exact position: it could be in the river, it could be on the bank. Whatever the case, its focus is completely riverine, the creature staring hard at the water below it, waiting silently for its cue to strike.

In the last hundred years there have been many developments when it comes to our rivers and the Otters that frequent them, transformations wrought by our ever-changing world. Some of those changes have occurred in the cast of characters with which the Otter shares the rivers; some of these characters have exited stage left, whilst others have entered unexpectedly, either helped by agents or by their own prowess. One character in particular has made quite a splash, returning to centre stage in an increasing number of modern-day riverine productions.

To exit stage left in the theatrical sense is to leave the stage in a quiet and undramatic fashion. When we lose species from an area, this is often how they go – there is usually no big dramatic finale, and instead they just quietly disappear, unnoticed as they go. It can be some time before we finally realise they are no longer present, and for those of us that take our seats after their exit has occurred, we often do not even know they were there in the first place.

I had such a moment when I read *Tarka*, taking my seat so to speak nearly a hundred years after it was written. I was reading the part where a female Otter pauses to listen out for sounds of danger or even opportunity, and as she listens she hears the call of a Corncrake.[1] When I first read that passage, I have to admit that I was a little confused; after all, the Corncrake is not a bird you find in Devon. I even wondered to myself if it had been a bit of writer's licence on the part of the author, but then the reality dawned on me. The Corncrake is not a bird you find in Devon *any more*. It was here, and it should be here, but sadly it exited stage left and few if any people noticed exactly when it went. Today the vast majority of us do not even realise that it was part of Devon's riverine cast at all. I know I didn't.

Corncrakes are migrant birds, members of the rail family, which includes more familiar species such as the Moorhen. They winter in the southern half of Africa and return northwards to breed across much of Europe and Central Asia; or rather, they used to breed over much of Europe and Asia. Corncrakes like to breed in grassland or, as we tend to call it, hay meadows. The mechanisation of agriculture, and in particular the mechanisation of hay cutting, spelt doom for so many populations of this secretive bird right across its breeding range. The Two Rivers Country in Devon was no exception. In the 1892 book *Birds of Devon*,[2] the authors note how the nests and eggs of the bird are frequently 'mown out' by haymaking, yet they also report that the bird can still be encountered frequently enough.[3] The meadows that the Taw passes through in Mid and North Devon were often wet in nature, as were many such meadows right across the country. They were great habitat for birds such as the Corncrake as well as many other species, but the agricultural revolution and the mechanisation and drainage brought with it would lead to serious declines in this bird's distribution. As the fields and meadows they used were 'improved', as they were drained and as mechanisation became commonplace, so the hay cut occurred earlier and earlier, and as a result any birds that were attempting to nest in such fields were no longer going to succeed in rearing young.

Corncrake were already in serious decline in Britain a hundred years ago when *Tarka* was written. They were most definitely moving leftwards across the stage in Devon, but they were still present and breeding here, although by this time their numbers would have already been severely reduced. Williamson's passage about a calling bird being heard in late May describes a breeding bird, or at the very least, one that was attempting to breed. Corncrake were just about hanging on in the wet meadows of Mid Devon, and some Otters in the centre of the county in the 1920s would have indeed heard them calling. But it is not a calling bird they or I will hear along the Taw today, nor is it one I will hear elsewhere in Devon. I am unable to even hear their curious song along other rivers in England,[4] Wales and most of mainland Scotland, because breeding Corncrake are now, sadly, virtually restricted to just the Hebrides and Orkney off the west and north coasts of Scotland. The riverine stages of Britain are far emptier than they should be.

Shifting birds

My having no idea that the Corncrake should feature in the Otter riverscape of Devon, and indeed England, is a great example of something known as 'shifting baseline syndrome'. In the context of nature, this can be defined as a gradual change in the accepted norms for the condition of the natural environment. Basically, because the Corncrake has not bred along the Taw in my lifetime, I see their lack of presence as a totally normal thing; to me it is perfectly normal for them to be absent, but if I lived a hundred years ago, I would (I hope) have expected them to be present. The baseline for breeding bird species that should be present along the Taw has shifted in the last hundred years, and as wildlife becomes increasingly depleted in this country, so those baselines are going to shift yet further.

There is another bird mentioned in the text of *Tarka* that provides another example of this syndrome of shifting baselines: the Turtle Dove. The Turtle Dove has, in my opinion, one of the most beautiful songs in the avian world, a soft, highly distinctive, purring sound that seems to drift on the air. It is a bird that I know well, having monitored them when I was a ranger working in forests in various parts of the country. Like *Tarka* in the book, I have often heard it singing in Devon, including alongside the Taw. Unfortunately, my monitoring of their numbers in Devon's forests reflected what was happening to them right across the country – each year the number of birds I heard would diminish until, in many places, there were none left to hear at all.

For me, the Turtle Dove is a bird that is missing (and greatly missed) from the riverscapes and broader countryside of Mid and North Devon. My baseline is set by my past experience of this bird, but for other, younger naturalists, the Turtle Dove is simply a bird that does not occur there. Their baseline is zero Turtle Doves, and therefore zero Turtle Doves is their normality. Shifting baseline syndrome is truly an insidious thing; it normalises our nature-depleted countryside without us realising it.

One hundred years ago, the Turtle Dove and its beautifully evocative song would have been familiar to people all over the southern half of Britain (although some birds did breed further north, the main breeding population was found south of an imaginary line from the Humber to the Dee). Its inclusion, like that of the Corncrake, in the story of *Tarka* was not an attempt by the author to embellish the soundscape of the narrative but was a reflection of reality, of the baseline of the time. They were mentioned simply because they were there. They have been lost in the intervening hundred years. The Taw's Otters hear them no more.

The Turtle Dove hasn't been completely lost from Britain, but its range has shrunk massively and its numbers are now worryingly low. There has been a 97% reduction in the bird's numbers in Britain since 1994. In other words, only 3% of the population of these beautiful birds that existed just over 30 years ago in Britain still exists here today. Their exit stage left is almost complete.

Shifting baselines can go both ways, though, and there are birds that have expanded their range since the days of *Tarka*, birds the fictional Otter would not have encountered. Otters today find themselves sharing British river systems with species that were simply not here a century ago, and one of those is another species of dove.

The Collared Dove is today a very familiar, and typically overlooked, bird across virtually all of the UK, from suburban gardens to remote rural farms. In 1927, however, it was a bird that had never been seen on these shores, and nor was it likely to appear here either – the nearest place you could find these gentle-looking doves was in Turkey. One hundred years later, though, and the baseline for Collared Dove in Britain has shifted dramatically. There are now around one million pairs of Collared Dove here, including along much of the River Taw. Today's Otters of the Taw will be very familiar with the bird's distinctive cooing call.

When a new species turns up in a country in which it has not been recorded before, and then goes on to rapidly form a large population, we tend to think of it as being an alien invasive species – a species introduced by humans that has now 'gotten out of control'. But the Collared Dove was not introduced to the UK by humans; this is a bird that introduced itself.

Just after 1930, three years after *Tarka* was published, the Collared Dove started to expand its range north and west, and it did so at remarkable speed. The exact reasons behind this sudden range expansion are unknown, but the bird has always been comfortable breeding in close proximity to humans, and it will readily take advantage of any feeding opportunities that we create for it. From the 1930s onwards, agriculture was changing in Europe, and perhaps as a result of these changes, it is likely that feeding opportunities for the bird increased. Whatever the exact reasons, the bird's range expansion was nothing short of remarkable. It had reached Germany by 1945, France by 1952 and Norway by 1955. But it did not stop there. If European domination was its goal, it achieved it by the 1970s, colonising the western edge of the continent in Portugal in 1974. It had even crossed vast seas to reach Iceland in 1971. In just 40 years the Collared Dove had completely colonised the European continent, increasing the size of its range by a staggering 2.5 million square kilometres in the process.

It was in 1955 that the first Collared Dove turned up in Britain, in Cromer in North Norfolk to be precise. But it was not just one bird, it was two, and not only did they first appear in 1955, they also first bred successfully that year too. A year before our Otter population would begin its calamitous collapse, the Collared Dove had begun its colonisation of Britain; by the time that Otters recovered and had returned to their rightful place in all of England's counties, the Collared Dove was a very common bird. So much so that the baseline had again shifted for birdwatchers – in the mid-1950s this bird's arrival would have created huge interest in that community, but today, many birdwatchers do not even give them a second glance.

Of all the birds associated with rivers and waterways in Britain, the Grey Heron is certainly one of the most iconic, and this statuesque bird gets many a mention in Williamson's book.[5] They are as much a fixture in the novel as they are in the rivers of today, but whilst they are written about by Williamson, their relatives the egrets are not. You could perhaps say that a hundred years ago, egrets were too few to mention.

There were no Little Egrets on the Taw in Tarka's day, but now they are not only found on it but also breed on it, as they do on many rivers across the country. The pure white plumage of these small members of the heron family dot the banks and the shallows of our watercourses and has become a familiar sight for anyone who spends time close to rivers. Two other species of egret have subsequently joined them, although in lesser number and mainly, for now at least, in the winter and during passage times (although both have now bred in Britain). They are the even more diminutive Cattle Egret and the much bigger, Grey Heron-sized Great White Egret.

However, unlike the Cattle and Great White, the Little Egret is a species of bird that Otters on the Taw (and on many other river systems across Britain) will have previously encountered. They are another example of shifting baseline syndrome, for Little Egrets are not newcomers to the British riverine stage; they exited stage left in the past before entering stage right once again. The Little Egret is a bird that was once found across the country, but records of its actual distribution, numbers and status (i.e. whether it was resident, a winterer, etc.) are unavailable, and therefore evidence for its existence in Britain is based on other records. Fortunately for those interested in the historical distribution of this species, but unfortunately for the species itself, the Little Egret was once considered good to eat; indeed, it was a bit of a gourmet bird. As a result of its perceived tastiness there are many records of its inclusion in various feasts and celebrations – these were birds that were on high-class menus. One example of this is the 1465 inauguration of the new Archbishop of York. As part of the investiture a massive banquet was held, and on the menu for this celebratory feast were 1,000 Little Egrets. That's a lot of birds.[6]

Little Egrets also appeared in great numbers on the menu for the coronation feast of Henry VI, as well as on many more, but of course this level of killing was in no way sustainable and inevitably their numbers in Britain dropped drastically. This decline in population was even noted with disappointment by the very people who were killing them for the table. In the middle of the 1500s the official Purveyor to the King's Mouth (now there's a job title) documented that he had to send further south for Egrets as they had now become scarce in the north of the country. The birds were being overexploited and they were dwindling fast as a result, but what finally did for them as a British species was not their flavour but a change in the climate. By the end of the 1500s Britain was entering the Little Ice Age, a period of over 200 years when the weather turned substantially colder across much of the Northern Hemisphere. By 1608 the River Thames in London was hosting its first frost fair, the river frozen solid enough for it to host a large public festival; during the winter of 1683–4 the Thames in London was frozen solid for two months with ice depths of almost 30 cm recorded.

Little Egrets are birds of more temperate climes; they are not lovers of icy conditions and this sudden and prolonged change in the temperature would have inevitably driven them further south into mainland Europe. This climatic change, coupled with our own insatiable appetite for them, meant that the Little Egret disappeared from Britain. Occasional individuals did turn up, but these would have been vagrants, lost birds wandering north from Europe. One was actually recorded on the river Axe in Devon in 1930, just three years after *Tarka* was published, but it was an exception rather than the norm.

Little Egrets continued to be hunted in mainland Europe, but this time it was more for their feathers than their flesh. During the Victorian era the bird's plumes became a must-have fashion accessory for the well-dressed lady, a fad that almost drove the Little Egret to European extinction, wiping them out from the north-west of the continent and leaving only low numbers hanging on in more remote parts of the far south. We in Britain, of course, played a significant part in this decline – we might not have had our own Little Egrets, but we had a huge demand for their feathers nonetheless. London imported vast numbers of Egret skins, replete with the desired feathers, during the 1880s. In just three months in 1885 over three quarters of a million Little Egret skins were sold in the London markets, and in 1888 one London-based dealer alone sold over one million skins.

Thankfully, like all fashions, tastes change. Given this waning of enthusiasm, combined with increasing public outcry and ensuing new legislation, the trade stopped. Little Egrets were no longer routinely hunted for flesh or feather and this allowed them to slowly recover numbers and ground; hunting and climate change drove them from Britain, and then the end of hunting and climate change allowed them to come back.[7] In the late 1980s they started to arrive in Devon again in increasing numbers, causing great excitement within the birdwatching community, and by the mid-1990s they were breeding in the county again. Due to shifting baseline syndrome, the birders that were so eager to see the species did so thinking that it was a new bird for the county and country, but it was not – it was a returning one.

The Little Egret is not the only fish-eating bird that suffered badly at the hands of humans, and as a result was missing from our riverscapes a hundred years ago. Like the Egret another bird has also made a return, although it is as yet not a complete one. Little Egrets are very beautiful birds and watching them stalk the shallows, waiting to lunge forward to catch a fish, is a fantastic sight. But it is not as spectacular as watching an Osprey smashing through the water's surface, talons first, before re-emerging from the refractive explosion of water droplets, carrying a fish tightly gripped in its highly specialised feet.

Ospreys, like so many birds of prey, suffered greatly from persecution for centuries. Their hooked beaks and long sharp talons were always going to encourage many to view them with suspicion, and the fact that they so blatantly took fish to feed on sadly sealed its fate as yet another vermin species that needed eradication. We tend to think of Ospreys as being birds of Scotland, catching fish on remote lochs and nesting in tall pines. But that is yet another case of shifting baseline syndrome. Ospreys are birds of Britain, all of it, including the Two Rivers Country.

Ospreys are migrants in Britain – they begin to arrive here in March to breed, before departing back to Africa at the end of the summer. In the 1500s they would have been widespread breeders, but in Catholic Britain they were seen as competition for fish, especially in overstocked fish ponds used by monks for their Friday fish suppers. Otters are attracted to these ponds and so are Ospreys, but unlike Otters, which would have raided these large ponds under the cover of darkness, the Ospreys did so in full daylight, typically making very noticeable large splashes as they did so. There would have been no tolerance towards their piscivorous activities.

Their extermination from much of England was methodical, even after the country's conversion to Protestantism. The last record for them trying to breed in southern England was from Monksilver on the eastern edge of Exmoor in Somerset in 1847, when a gamekeeper shot the pair as they built the nest. As is often the case, there is no official last date for them breeding in England, but it is likely that it was only a few years after this doomed attempt. In the 1500s and 1600s Ospreys were regular breeders in many parts of Devon, especially favouring estuary and coastal sites for nesting, including the Taw estuary. But the nests of this magnificent raptor are very obvious and easy to locate; as the war on wildlife gathered pace following the passing of the Tudor Vermin Acts, so their nests were quickly found and the birds that built them were killed. Ospreys rapidly disappeared as breeding birds. The last record I can find for a mainland Devon breeding site was on the cliffs of Beer on the East Devon coast, where it was recorded as a regular breeder in the mid-1700s. The last breeding pair for Devon as a whole was on the island of Lundy in the Bristol Channel. They bred regularly there until 1838; the island's location in the middle of the fast-flowing channel offered the birds some protection from the onslaught, but in that year the male was shot by one of the Channel pilots for no reason other than he could. The birds never bred there again.

Even though the breeding numbers were dwindling fast in the county, Devon was still regarded as the best place in England for them during the Georgian era. The ornithologist George Montagu stated in his *Ornithological Dictionary*, published in 1802, that 'the Osprey is more common in Devonshire than in any other part of the kingdom'.[8] The birds that were being seen in Devon were not staying in the county, though. They were just passing through on their migration, both in the spring when they were heading northwards and again in the late summer when they were heading southwards. There are numerous records for Ospreys on passage along the river systems of Devon in the first three quarters of the 1800s, but when looked at closely they underscore the reality these birds were facing across the whole country. Words such as 'procured' and 'collected' in amongst the dates and place names of the records demonstrate the fate of the birds recorded: they were being killed.

The killing of the birds on passage and the killing of them at their breeding sites further north inevitably led to a steep decline in their population. Even the supposed finest place to see them in the kingdom soon became a place where they were no longer seen at all. In the aforementioned 1892 book *Birds of Devon*, the authors make note of the last record for them, when 'Quite a flight seems to have visited Devonshire in the autumn of 1875, when Ospreys were killed on the Teign, Dart, Avon, Tamar and Taw.'[9] That record was followed up by a somewhat telling sentence: 'Since that year none have occurred in the county within our knowledge.' You would hope that they realised why.

We systematically exterminated the Osprey from England, Wales and finally Scotland a few years into the twentieth century. It was in 1916 that the last pair attempted to breed, at Loch Loyne in the Scottish Highlands. But as with the Little Egret, when our relentless persecution of them stopped, they too made a comeback. From 1920 to 1948 a total of 22 records for wandering Ospreys exist for Devon, with the majority from the Exe estuary, still a great place to see these birds on passage today. The river with the second-highest total, six in 28 years, was Tarka's Taw. As *Tarka* was being published the Osprey was still extinct as a breeding species in Britain; it is not mentioned in the book. Six records over nearly three decades for the Taw make it no more than a rare vagrant to the river system, and in the 1920s they were not considered part of the Taw's riverscape. But a hundred years is enough time for things to change.

Ospreys officially returned to Britain as a breeding species in 1954 when the famous pair bred at Loch Garten in Scotland; however, there is other evidence to suggest that they perhaps had been breeding in the remote Scottish Highlands since at least 1947. Those first few years, though, were a struggle – intensive protection had to be installed around nest sites and 24-hour wardening schemes were used, not as a defence against people wanting to kill them for hunting fish, but against a selfish, miniscule and in my opinion somewhat pathetic number of people who wanted to steal their eggs so they could add them to their collection in a dark dusty drawer.

The egg collectors had some successes causing some of those early nests to fail, but generally the Ospreys were able to breed and produce young; and as more of those young were fledged, so the potential for the very small and fragile breeding population to grow increased. But, as with the Otter, the mid-1950s were about to herald in a time of yet more death and destruction. The pesticides that decimated the Otters from 1956 onwards also had an impact on the Osprey. Otters accumulated their dose of pesticides from their diet, the same diet of fish that the Osprey also eats. As the Osprey's breeding sites were in the remoter parts of Scotland, the level of toxicity from these agricultural pesticides was relatively low when compared to other parts of the country – but of course these birds migrated through England (and often Devon), where they continued to hunt and eat fish.

It is not clear whether the returning British breeders were adversely affected by the pesticides of the 1950s and 1960s; certainly Ospreys in Europe were impacted by the effects of these pesticides on their breeding success and as a result the overall European population took a long time to recover from its suppressed numbers, even as direct persecution itself eased. However, today their numbers are increasing once again and their British range is increasing too, thanks to both the natural spread of the breeding population and a few reintroduction projects, and they are no longer just breeding in Scotland. Ospreys now breed in Wales and England too, places where they belong. Overall, there are between 250 and 300 pairs of this fantastic fish-eating raptor breeding in Britain today.

They have not recolonised Devon as a breeding bird yet – I hope it will not be too long until they do – but the county is still a great place to see them on passage when they use the rivers and estuaries to stock up on food before continuing their journey. The Taw is still a favoured spot, especially the estuary between Barnstaple and the sea, but I have also seen them further upstream, slowly flying over the main road that snakes along the section of the Taw shortly after the Yeo has met and joined it. Ospreys are back again in Williamson's Two Rivers Country. The Otters, of course, will be indifferent to their return.

As well as the Osprey and Little Egret, our Otters today share their riverine habitat with another bird that also was not mentioned by Williamson in *Tarka*. This one, though, is not so easy to see as the gloriously white Little Egret, nor is it as spectacular as an Osprey. When present, however, it is easy to find as this is a bird that quite literally calls attention to itself. The Cetti's Warbler is a small brown bird that prefers to spend its time skulking and hiding in dense patches of wetland vegetation. This habit makes them difficult to see well, although with a bit of patience they do usually show themselves. It is their distinctive and proportionally very loud song that reveals their presence long before they are actually seen. It is a song that Williamson would have most certainly mentioned a hundred years ago if they had been present in the Two Rivers Country, but they weren't. The Cetti's Warbler is one of Britain's newest resident birds; it was first recorded as breeding in Britain, in the county of Kent, in 1972. Since then, it has expanded its range greatly, and although more often associated with the south of England, it is heading further north each year with birds now breeding in northern Wales, northern England and, since 2023, southern Scotland.

The Cetti's Warbler is now a bird of the Taw, particularly where the river becomes an estuary towards the end of its journey. Its explosive song is now a reliable fixture on this section of the waterway and it will be a sound familiar to any Otters travelling this stretch.[10]

The Cetti's Warbler, Osprey and all three species of Egret have got here under their own steam. They have colonised (or recolonised in the cases of the Little Egret and Osprey) Great Britain naturally, part of a northerly spread of many birds that has been associated with the increased temperatures of climate change, the creation of wetland reserves and better habitat quality in general. But not all newcomers to our river systems in the last hundred years have arrived unaided; others are only here because we humans have facilitated their arrival, either deliberately or accidentally. And when we do that, things often turn out badly.

American arrivals

A natural, and often very important, part of the Otter's diet is provided by crayfish, miniature riverine versions of the ocean-dwelling lobster. If you go hunting for Otter spraints as I often do, you may well be familiar with their distinctive broken remains. In Britain we have our own native species of these crustaceans, the White-clawed Crayfish. It should be the only species of crayfish found here, and it was a hundred years ago, when it would have been a regular prey item for Otters across most of the country, including on the Taw. But now, our native species is no longer alone; now it is just one of a total of seven species of crayfish that can be found in our rivers and water-ways, with the other six arriving here at various times since the 1970s. One of these new arrivals has become particularly widespread and has had a devastating impact on the distribu-tion of our native White-clawed Crayfish.

The Signal Crayfish is a North American species; it should not be anywhere near Europe, let alone the British Isles, but despite this it was deliberately introduced to Britain in 1976. It is bigger than our native species of crayfish and therefore it can outcompete it ecologically, and whilst this was always going to be an issue for the White-clawed Crayfish, its American cousin also brought with it something called crayfish plague. As that name suggests, this was not a good thing.[11] Crayfish plague was first recorded in Britain in 1981 and is lethal to our native White-clawed Crayfish, which has no natural resistance to the disease, with infected animals usually dying within a few weeks of contracting it. In 1976 we introduced a crayfish species to our waterways that spreads a disease that does not affect it – but unfortunately it is a disease that does disastrously affect other species. As a result of our actions, the Signal Crayfish has outcompeted and ultimately decimated our native crayfish populations.

The White-clawed Crayfish has been wiped out from many parts of its British range; the combination of a larger competitor, a virulent disease and, let us not forget, increased pollution of our waterways has simply been too much for it. As a result of our actions there are no White-clawed Crayfish on the Taw any more for its Otters to feed on, but there are now plenty of Signal Crayfish. From the Otter's point of view this change in species probably has not had any impact. After all, the crayfish food source is still there, and the 'new' species is actually bigger than the 'old' one and therefore makes more of a meal. But from our point of view it is yet another important lesson for us humans about why we should always guard against introducing non-native species. Unfortunately, the arrival of the non-native crayfish in our river systems was not the first lesson we have had on the subject.

A couple of years after *Tarka the Otter* was first published in 1927, a distant relative of the Otter was brought to Britain. The American Mink is a much smaller species of mustelid compared to the Otter,[12] though it does share the same semi-aquatic lifestyle. But this North American species was not transported to Britain to be released into our waterways; instead it was brought here to be kept in the confines of small cages in mink farms. Farms whose only output was the fur and skins of the caged animals, products that would supply the fashion industry's predilection for fur coats and grotesque stoles.

Inevitably, it was not long before some of these American Mink escaped from their confines – and who can blame them for trying to flee the abject misery of a fur farm? The housing was often inadequate on the early fur farms and in the typically resourceful manner of all mustelids, many American Mink made light work of their confinement, escaping into the wider environment. Others were released en masse by animal rights activists, especially in the latter half of the twentieth century when public awareness and distaste for the fur industry grew substantially. Whether they escaped themselves or were released by humans, it did not take very long for populations of these carnivores to become established in various parts of Britain, with the first self-sustaining populations of American Mink being recorded on the River Teign in Devon in the 1950s. By the early 1960s, several other self-sustaining populations of American Mink had become established in a wide variety of locations across the country. By the 1990s American Mink could be found virtually throughout the whole of Great Britain.

The American Mink became firmly established in our countryside just as the Otter's decline was being noticed, and there were some that wanted to suggest that the American mustelid played a part in the decline of the Otter. But that wasn't the case, and those that tried to promulgate the theory were ignoring some pretty basic information. Firstly, they were ignoring the real cause of the Otter's decline: our use of lethal chemicals in farming. But they were also ignoring the fact that male and female Otters are about three times the size of male and female Mink respectively, and that male Otters weigh on average seven and a half times more than a male Mink whilst female Otters weigh as much as ten times more than a female Mink. The size disparity between these two predator species is huge; to blame the Mink for the Otter's demise is the equivalent of blaming the European Lynx for the extirpation of the Lion from southern Europe.[13] Otters do not view Mink as something to be scared of, but they might view them as something to eat.

As Otters have returned to their rightful river haunts, so they have come across established populations of American Mink, a species they had never interacted with before. Research has demonstrated that the far larger Otter is bad news for the American Mink – studies show that as Otters returned to rivers, the density of American Mink dropped drastically. Other studies have shown that American Mink respond negatively to the odour of Otters, completely avoiding places that smell of the much larger mustelid. Both species use their spraints as a form of communication, and it appears that an Otter's spraint is a communication of which the American Mink takes heed. The Otter is easily able to outcompete the American Mink in the pursuance of aquatic prey items, and the introduced species simply cannot cope with this competition. It has been observed in other studies that in the presence of Otter, American Mink will switch to a more terrestrial prey base; in other words, they stop hunting for food in rivers. This is likely due to Otters not only outcompeting the much smaller predator, but also behaving aggressively towards them. Throughout the natural world, larger predators often dominate smaller ones, driving them from their territories and even sometimes killing them. The presence of American Mink remains in both Otter spraints and the stomach contents of Otter road casualties shows that not only will they kill them, but they will also utilise them as a protein source.

American Mink are declining in Britain at the same time as Otters have returned to every part of it. Recent national surveys of Otters have shown that the decline is steep and continuing.[14] In the English National Otter Survey for the year 2000, of the 3,300 sites surveyed 729 had positive signs of American Mink presence. In the survey conducted in 2010, of the 3,300 sites surveyed the number with positive signs had dropped to 510. In the latest survey carried out in 2024, this number (again out of 3,300 sites in total) had dropped even further to just 91. In the first 24 years of the millennium, the American Mink has declined from being present in 22% of sites surveyed to just over 2%; it is a huge decline in numbers and raises the possibility that the American Mink's days in Britain are numbered.[15]

This decline is most likely down to targeted culling programmes of the American Mink by humans, but we must not ignore the fact that the Otter is also reasserting its right to be the top mammalian predator of our waterways, and it evidently does not appreciate the smaller newcomers. We often think of large predators – and we should not forget that the Otter is a large riverine predator in Europe – as regulating the populations of their prey, but they also regulate the populations of smaller predators too; it is part of their ecological role. American Mink have caused huge problems for many species in our river systems since they became established in the wild. Their presence, combined with the severe degradation of our riverine habitats, has led to local extinctions of some species such as the Water Vole. American Mink have had a negative impact on our wildlife, but they played no role whatsoever in the drastic decline of the Otter in Britain from the 1950s onwards – we did that all by ourselves.

The American Mink is not the only mammalian arrival to our waterways since the days of Tarka. An even larger species has made its presence known, and it is one that Otters today are increasingly likely to encounter in many parts of the country. But this is not a new species for the Otter or our countryside; instead it is a returning one.

The three strangers

As we have already seen with the likes of the Corncrake, the depletion of Britain's wildlife due to our actions over the last hundred years has been both vast and rapid in its magnitude. In short it has been devastating. But this degradation, this depletion, of our fauna, our flora and the habitats in which they exist started long before the 1920s.

Silent Fields by Roger Lovegrove is a brilliant book, even if its content is somewhat depressing to read.[16] It chronicles our assault on wildlife since the Tudor times, its meticulous research detailing our relentless war on a variety of species from Kingfishers to Kites and Bullfinches to Badgers. However, the sad truth is we had already lost many species before the start of the Tudor dynasty in 1485, when the soon-to-be Henry VII defeated the incumbent Richard III at the Battle of Bosworth Field.

Amongst those pre-Tudor losses were some iconic mammals, ones that long ago exited stage left. The Brown Bear, that magnificent beast of an animal, had most probably been hunted to extinction in the British Isles before the Romans arrived here in 43 CE – although of course, the actual date these very large predators became extinct in Britain is something we will never know.

The Northern Lynx, a beautiful feline predator and the natural regulator of our populations of Red Fox and Roe Deer, hung on for a bit longer than the Brown Bear, outlasting the Romans and their empire and probably still surviving here until as late as the seventh century. Then it too was gone, but again, it is difficult to be precise with the dates. It is a species that is currently the subject of much debate as to whether it should be reintroduced to Britain, with Scotland looking to be the most likely candidate for any release scheme.[17]

And then there is probably the most iconic lost native British mammal of them all, the Wolf. It vanished forever from these islands in the seventeenth century, when it lost its last foothold in the Scottish Highlands. There is a bit more documentation relating to the demise of the Wolf than for the other two species, but the vast majority of it is apocryphal nonsense, with all sorts of outlandish claims made about which person took the 'glory' of killing our last large native carnivore.[18]

All three of these mammals are ones that British Otters would have lived alongside for thousands of years, even if they were all large predators that Otters would have probably tried to avoid, using their semi-aquatic lifestyle to maintain a safe distance. But there was one large mammal that shared that semi-aquatic lifestyle, and one that Otters would have lived very closely alongside indeed. The Eurasian Beaver, a mammal that we humans also hunted to extinction in Britain.

The return of the native

As with the other extirpated mammals mentioned above, no one really knows for sure when the last native Beaver in Britain left the stage. It is likely to have been in the sixteenth century during the Tudor era, although it is possible it managed to hang on for a bit longer; all we can say for certain is that it went extinct several centuries ago, in an era when no one either really noticed or cared. The Eurasian Beaver was once widespread across the country, with strongholds in the wetlands of East Anglia and Somerset amongst others. But by the time the fictional Tarka slipped into the headwaters of the River Taw in the 1920s, it had been extinct in Britain for a very long time.

Unfortunately for this large rodent, we humans valued the mammal enormously, but this anthropic value sadly had nothing to do with its vital role in the ecology of our rivers and wetlands. Instead, we valued the Beaver for the products we could create from it. To us, the Beaver was the source and supplier of three important commodities: meat, fur and something called castoreum.

The first written mention of the word 'castoreum' in the English language was in a book entitled *On the Properties of Things* in the late fourteenth century.[19] The Beaver is described in the book as being a 'helpyth ayenst many syknesses' – in other words, it was a medieval medicinal cure-all, a panacea for the populus, or a beaver for a fever if you like. But this supposed medieval miracle cure is, in reality, nothing more than a yellowish exudation from specialist gland-like structures that all mature Beavers have close to their anus. It is this that we call castoreum. Beavers use their castoreum, mixed with their urine, to leave a detailed scent mark for other Beavers to 'read', in much the same way that Otters use their spraints to communicate information to other Otters. We humans, however, used their castoreum to try to treat a wide range of afflictions, everything from mild headaches to the rather more serious Bubonic plague. It should be noted that the Beaver's usage of castoreum was, and still is, somewhat more successful than our own.

Unsurprisingly, castoreum has no scientifically proven medicinal usages; of course it hasn't, for even though it didn't come from a duck, castoreum was never anything but a quack cure. However, that did not stop people from trying to use it to alleviate any affliction they could think of, and our use of it for such purposes had been going on for a very long time before *The Properties of Things* first went to print. The Romans were incredibly fond of using castoreum, and they no doubt brought that trend with them when they arrived in Britain, a trend that continued long after the Roman Empire had itself crumbled away into history. Unfortunately for the Beaver, we obtained the castoreum we thought we so desperately required by killing the animal, and by the time the word 'castoreum' first appeared in written English, its source, the Beaver, was virtually extinct across most of Britain.

Our use of castoreum was not just limited to quack medical cures; it has also been used in the manufacture of perfume and in food flavouring (vanilla, if you are wondering). But by the time these later uses for it came along, the British Beaver was no more.

The Beaver was also a source of meat; a large plump rodent would no doubt have been a tempting target for protein-seeking humans who had no shops to head to when looking for something to sustain themselves. Its usefulness as a food source was compounded by the fact that, according to Catholicism, the flesh of the Beaver, due to the animal's aquatic habits, was considered akin to fish and therefore it could be eaten during religious fasting periods such as Lent, when the eating of non-fish meat was forbidden.

But we did not just think of the Beaver as being a source of supposed medicines or non-mammalian meat; we also had our eyes on the wrapping that these products came in. Its fur. Beaver fur has long been used by humans to make clothing, everything from rudimentary coats to top-of-the range hats. It is dense stuff, with as much as 23,000 hairs per cm^2 – not as dense as Otter fur (see Chapter 1) but nevertheless, still very insulating when used in the manufacture of clothing. Beaver fur also has another very useful property when it comes to the making of clothes. The fur of a Beaver differs from Otter fur, and from any other mammal's fur in fact, in its unique structure. The individual hairs that make up the Beaver's fur coat are barbed, allowing them to interlock perfectly with one another, creating not just a warm coat but a waterproof one too, crucial for an animal that spends so much of its time in the water. Otter fur is denser than the Beaver's, but it is not waterproof. Otters need to regularly come out of the water to dry off, but Beavers do not. It is this warm and waterproof property that made the fur of a Beaver such a highly prized commodity in the human world.

It was something of a triple whammy for the Beaver in Britain (and elsewhere within its range) – the unwitting mammal was the source of three human-valued products that could only be obtained by killing the animal,[20] and it was our greed for these products that sped this unfortunate creature rapidly to extinction as a British species. For many centuries, the Otters of Britain have had no interaction at all with their fellow native riverine mammal.

But that is now changing and it is changing rapidly. Tarka would not have come across a Beaver as he swam down the Taw a hundred years ago, but the modern-day real Otter would. For the Taw is beginning to be resettled by these brilliant mammals, and it is not the only river system in the country that has seen them return. Beavers are today loaded with many appellations; they are called eco-engineers, habitat creators, a cornerstone species, but more importantly than all of that, they are native British mammals that belong here. British Otters belong with British Beavers. The return of the Beaver to Britain has long been overdue.

The reintroduction of the Beaver to Britain had been discussed for many years; there had been a wide variety of project groups and committees set up to look at the feasibility of bringing back this native mammal. It was a typically long, drawn-out process, and one that met with much opposition in some quarters, but eventually it was announced that a small number of Beaver would be released into a heavily controlled area in Knapdale Forest in the west of Scotland. These animals would then be carefully monitored by a team of experts, their actions subject to much study and analysis, before any decisions on the future of Beavers in Britain would be taken. It was British Beaver bureaucracy at its best.

The date for the start of the Knapdale trial was 2009. But the Beavers had other ideas, or to be more accurate, persons unknown had other ideas, far less bureaucratic ones… In 2008, a year before the official release project was due to start, wild-living Beavers were discovered on the beautiful, and for this book appropriately named, River Otter in East Devon. The river is one I know well; I grew up in between the River Exe and the River Otter and spent many a happy hour splashing about in its waters as a youngster. So when the whispers started and then the news officially broke, I was thrilled to think that this river was the first in Britain for many centuries to have Beavers living wild in it again. Their sudden appearance, though, prompted an obvious question. How did they get there? These Beavers were not part of an undiscovered population that had somehow survived undetected for hundreds of years, nor were they a result of an escape from a zoo, and of course, the animals had not naturally recolonised from mainland Europe, somehow sneaking across the English Channel under the cloak of darkness. The answer was very simple: the Beavers were back in the River Otter because someone had deliberately put them there.

In 2004, four years before their appearance on the River Otter in Britain, wild-living Beavers were suddenly found to be in the Ebro river system in northern Spain, the first time they had been recorded anywhere in the Iberian Peninsula since 1583. Again, questions were asked as to their origin, culminating in a genetic study being carried out. The results of that study determined that the Beavers had originated from fur farms in Bavaria in south-eastern Germany; somehow, persons unknown had released them from their confinement, transported them across Europe and then released them again in Spain. Whether this was the inspiration for what consequently happened on the River Otter in Devon is just conjecture on my part – although somewhat interestingly, genetic analysis has shown that the River Otter Beavers are also from the Bavaria area of Germany, which is quite some coincidence. The origins of the Otter's Beavers may be uncertain, but the one thing known for sure is that they are finally back.[21]

Currently, the actual full extent of the Beaver's reclamation of its rightful place within our river systems is still somewhat unclear; you could say the waters are as murky as the animal's origins. In addition to the bombed Beavers, there are a number of officially licensed projects, including Knapdale, where Beavers are being studied within fenced sections of river systems across the country. There are also an ever-increasing number of rivers where free-living populations are being discovered and publicly acknowledged. It varies dramatically on who you ask, but a ballpark figure of at least a couple of dozen river systems now containing wild-living Beavers seems to be widely agreed, although in truth that number is bound to be an underestimate and probably quite a big one too.

One of the rivers where the Beaver is confirmed to be present, despite the fact that apparently nobody officially knows how they got there, is the Little Dart River in Devon. Although its name might suggest a connection with Dartmoor, it does not rise or flow through this iconic Devonian upland; instead it rises in the north of the county near a village called Rackenford, before flowing west to meet the River Taw near the town of Chulmleigh. It is where the Little Dart meets the Taw that Beaver activity has been recorded, and it is here that the Otters of the Taw get to interact with Beavers once again.

But how do Otters and Beavers interact? The population of Beavers on the River Otter became the focus of much discussion once they had been discovered, and at one stage the government of the day, no doubt under pressure from certain quarters, even talked about eradicating them. Thanks to much lobbying by groups such as the Devon Wildlife Trust, as well as a lot of public support, common sense prevailed. A long-term study, called the River Otter Beaver Trial, led by the University of Exeter and the Devon Wildlife Trust, was instigated and duly carried out. The trial looked at all aspects of the Beavers' presence in the River Otter and its tributaries, and one of those aspects included how they interacted with the Otters of the Otter.

One of the main concerns around unexpected populations of Beavers suddenly appearing is that, unlike in officially sanctioned reintroduction projects, there is no record of rigorous health checks on the animals involved.[22] This might seem a superfluous concern, an overly cautious bit of bureaucracy – but it isn't, there are risks. The biggest risk with rodents, of which the Beaver is one, is a particularly nasty type of tapeworm called *Echinococcus multilocularis*. This tapeworm parasite is not currently found in Britain, but the government is sufficiently worried about it to categorise it as a notifiable animal disease, meaning that if you suspect it is here, you must notify the appropriate government department or risk being prosecuted.

Tapeworms, like many parasites, have a complicated life cycle; this particular species is a parasite of canids, the dog family, with foxes being at particular risk of infection, but domestic dogs are also very prone to this tapeworm.[23] But this species of tapeworm doesn't just need dogs or foxes as a host, it also needs a rodent. The larval stage of the tapeworm develops in cysts within the bodies of rodents – usually voles, but Beavers are also infected in some parts of their range. The cysts are subsequently transferred to the canid species when it preys upon, or scavenges the carcass of, the rodent. If this tapeworm were present in a Beaver population in a river system where people regularly walk their dogs (as they indeed do along the River Otter), there would be a relatively high risk of this parasite spreading into the domestic dog population and potentially the human one too.[24]

The tapeworm risk was deemed serious enough for the government to require that the Beavers present on the River Otter were tested for it before the trial could commence. For this to happen, the Beavers had to be trapped and then taken away from the river to be kept in quarantine whilst the health screening took place. This happened in 2015, which meant that the river would once again, but this time hopefully only for a short period, lose its Beavers.

The availability of potential Otter holts within a river system can limit its carrying capacity for breeding Otters. As we have already discussed, Otters do not make their own breeding holts, but they are very able to adapt any potential holt opportunities they discover. On our modern-day rivers though, these opportunities are typically few and far between, unless of course that river system has Beavers.

The presence of Beavers in a river system raises the potential for holt sites exponentially. This is due partly to their ability to transform the riverscape, completely changing the habitat it provides. In Britain we have largely forgotten what a riverine system should look like. Our rivers may look bucolic, gently winding their way through pastoral landscapes, with the odd tree dotted here and there along its banks; this may look picturesque, but it is not in any way natural. Lowland rivers in Britain are artificial landscapes.[25] A river should not be a linear waterway cutting through an expanse of dry land. It should be a dynamic mix of wetland habitats.

I met up with Pete Burgess, Director of Nature Recovery for the Devon Wildlife Trust, to catch up on old times and to chat about the Beavers on the Otter, as he was heavily involved in the River Otter Beaver Trial. He summed up perfectly what it is that Beavers do: 'to put it simply, in our headwater streams Beavers take that picturesque scene and they make it exponentially more complex and dynamic; some might see that as Beavers making a mess, but in ecological terms they create an unrivalled diversity of habitats'. It is this creation of habitat that in turn provides food, cover and breeding sites for a whole host of species, including the Otter. As Pete says, 'Beavers make Otter heaven.'

But it is not just their habitat creation that boosts the number of potential holt sites for Otters to use. Beavers are also prolific builders, forming structures that we call lodges, places in which they reside and breed. They will have a main lodge, the focal point of their social activity, but they also create smaller 'bank' lodges within their territory that they use on a much more infrequent basis, in much the same way that Badgers have a main sett as well as numerous satellite ones.

These outliers of course do not pass unnoticed by Otters, who eagerly use them as ready-made holts. Like most successful animals, Otters are opportunistic and as such they are not going to pass up on the housing opportunities provided by Beavers and their lodges.

The Otters of the River Otter demonstrated this opportunism when the Beavers were removed for their health screening; as soon as they were gone an Otter immediately took up residence in the Beaver's main lodge, effectively turning it from a lodge to a holt. But as the Otter was settling into its new home, the Beavers were getting a clean bill of health. There were no tapeworms within them, and they had therefore been officially cleared for release and a return to their territory. Pete and the rest of the trial team were not sure what would happen once the Beavers returned to the river. They are large rodents, but Otters are large carnivores. The River Otter Beaver Trial team were playing witness to a natural relationship that had not been seen for centuries in Britain.

Any questions as to what would happen were very quickly resolved. As soon as the Beavers were released back into the river, they returned to their lodge and turfed the Otter out. There was no long, drawn-out conflict between the two species; the Beavers simply moved back in and the Otter moved out. Pete used the phrase 'mutual distrust' when describing how the two species interact. The Otters are wary of the large and powerful rodent (and no doubt its formidable tree-gnawing teeth), and the Beavers are also wary of the predator, but both generally tolerate each other's presence, with the trial team recording an active Otter breeding holt within 100 metres of an active Beaver breeding lodge. Beavers will not tolerate Otters too close to their breeding lodges though, and seem to actively push them out from the immediate area if small young are present. This is probably because the rodent perceives the carnivore as a potential predator of its small kits, and an Otter would certainly be able to deal with a Beaver kit if an opportunity presented itself. The River Otter Beaver Trial team has never found any evidence of this happening, although there is anecdotal data from elsewhere in Europe that suggests that it does occasionally occur.

But whilst there may be an air of mutual distrust between the two species, there are also mutual benefits in their coexistence. Pete told me how the habitat created by the Beavers forms a rich mosaic of pools and channels of water of differing depths; this variety of depth and flow is perfect for a whole host of fish species and their various life stages. Fish are greater in number and, importantly, weight in Beaver-created habitat. This leads to a healthier fish population and of course also to better food availability for the Otter (and other piscivores such as the Herons and the Egrets), which readily takes advantage of the easier hunting. It is well known, and is indeed a principal argument for having them back, that through their activities Beavers regulate the flow of rivers, reducing the risk of flooding and the damage this can wrought on our own communities. But this regulated flow also means the Otters are better able to exploit the habitat in times of increased rainfall. Without Beavers, our linear rivers can flow at alarming speed following periods of heavy rain, which in turn can force the Otters to move away from the narrow, hemmed-in watercourse. Otters are good swimmers, but a high river in fast flow can be a dangerous place even for an experienced adult. When an Otter is forced into moving from a section of river, this can not only bring it into conflict with other Otters holding territories elsewhere, but also brings an increased risk for the Otter as it attempts to circumnavigate the fast-flowing waters. But that is a risk we will examine in another chapter.

A river that has been Beavered creates opportunities for Otters to exploit the broad wetland habitat that surrounds the no-longer-linear flow. The water levels will still rise in times of high rainfall, but the flow is slowed; the presence of Beavers provides Otters with the opportunity to find places within the now complex riverine habitat where they are not at risk from being washed away, places where they are still able to easily find food. Thanks to Beavers they are unlikely to have to move to new places in times of flood, and can stay within the river system safely.

I also asked Pete if the Otters bring any benefits to the Beavers, to which he replied that the Otters act as an early warning system, using their acute senses to detect any threat. In Britain today this threat is most likely to be a dog off the lead, whereas in the past it could have been a pack of Wolves hunting along a watercourse, but the principle is the same. Beavers watch what Otters are doing and if the Otter detects a threat, the Beaver soon observes this and reacts accordingly.

It must have been fascinating and incredibly rewarding for Pete and the rest of the River Otter Beaver Trial team to see this long-evolved coexistence happening all over again. It should be no surprise to us that Beaver and Otter can coexist and that this relationship can have mutual benefits – after all, it has been happening for millennia, and it was only us humans and our actions that disrupted it. Otters, Beavers and the rivers which they both inhabit belong together.

As I was finishing the manuscript for this book, the British government announced that following the numerous Beaver study trials across the country, they have approved the release of wild Beavers in England.[26] Releases will be overseen by the government agency Natural England and will be done under licence; the first of these approved and legal releases happened within a week of the government's announcement, with two pairs of adult Beavers being released into the wild on a National Trust land holding in Purbeck, Dorset. The released Beavers had been translocated from Scotland where they had been captured as part of the Scottish government's strategy for dealing with conflicts that the Beaver's presence may cause for some landowners.

Beavers have been back in England for nearly two decades, and it is great news that after a long process their presence is now legally sanctioned.

A single family tree

It is not just the supporting cast of fauna on and in our waterways that has changed over the last century; the flora has also had its changes too. Many of these will have had little effect on the Otters, but they can be extremely noticeable to us. Probably the greatest example of this is Elm. English Elm trees were once a very common feature of the landscape in many parts of England, and would often be found growing alongside watercourses, their distinctive dendrite shape forming the backdrop to many a landscape painting. Perhaps their most famous depiction is in *The Hay Wain* by John Constable, an oil on canvas work that depicts the river Stour as it flows along the Suffolk–Essex border.[27]

That famous scene can still be viewed today, but the Elm trees that were there in 1821, when Constable completed the painting, cannot. They, like the vast majority of English Elms that dotted our landscapes, have long gone. But before we explore that, we should clarify what an English Elm actually is. Because for a start, they are not English.

Taxonomically, the group of trees that we call Elms, specifically those in the *Ulmus* genus, are a very confusing bunch. Until relatively recently it was thought that Britain was home to a number of different Elm species, trees with names such as the English Elm, the Huntingdon Elm and the Smooth-leaved Elm (and many more). But all of these are in fact the same tree species, the Field Elm. Or to be more accurate, they are all the same clone of Field Elm, a clone that we once called English Elm, but today is known more accurately as the Atinian Elm, a name that relates to its place of origin, Atinia in the central Italian region of Lazio.

Yes, English Elms are actually Italian – or, to be completely accurate, Roman. Britain does have a native species of Elm, but it is not the English Elm but the Wych Elm,[28] a different species altogether from the clone of Field Elm that we called English Elm among many other names. Our prolific naming of these trees is as confusing as their biology.

Field Elms don't (generally; nothing is cut and dried with this group of trees) reproduce by fertilised seed, but instead form clones of themselves by producing suckers that grow off the parent tree's roots and then spring up out of the ground nearby. These suckers can be cut off by us and planted into the ground where they will then form roots of their own, producing an identical clone of the tree from which they originated. Humans have long realised that they could take these cuttings to create new trees, and by Roman times it was standard practice in much of the tree's natural range. The Atinian Elm clone that we would later mistakenly call the English Elm is believed to have originated from a single tree growing in Atinia. Suckers from this tree were taken by the Roman agriculturalist Lucius Columella,[29] who subsequently introduced them to his vineyards in Cadiz in southern Spain where they were grown, as they were back in Lazio, as vine supports, effectively forming a living trellis. This trellis was multi-purpose, with its leaves being stripped off to provide cattle food, which also allowed increased light through to ripen the grapes they supported.

The suckers used to create these vine supports would in turn produce suckers themselves; these were then cut off and planted to support other vines both nearby and further afield, and so the cloned progeny of just one single tree spread rapidly over south-west Europe. It did not take very long for it to reach Britain either, with it being introduced here by the Romans at some point in the late first century CE.

The Roman Empire may have crumbled, but the use of Atinian Elm in agriculture continued apace after its fall, with cuttings of suckers being planted prolifically to provide leaf fodder for livestock and, later on, as a hedging plant that was widely used from 1604 in Britain following the start of the Inclosure Acts. All of these Elms, every single one of them, including the famous examples painted by John Constable, were genetically the same tree: a near-2,000-year-old clone that had originated from just one tree growing in central Italy. This meant that in effect there would have been millions of identical trees growing all across Britain and western Europe.

Whilst that may be remarkable to think of, it is also very risky. Low genetic diversity equals low natural resistance to pathogens and diseases, and with the genetic diversity of the Atinian Elms being roughly zero, it was only a matter of time before something came along that they would not be able to resist. That something was Dutch Elm Disease.

Dutch Elm Disease is a fungal pathogen that is spread by small insects called bark beetles.[30] These insects are tiny and would otherwise be insignificant to the vast majority of life, if it were not for their unwitting role as the vector of the fungus. Dutch scientists first recorded the fungus in the 1920s, and soon afterwards it was recorded in Britain, discovered by the Forestry Commission's Dr Tom Pearce in Hertfordshire. The strain of fungus in pre-Second World War Britain, however, had only a minor impact on the trees here and by 1940 the outbreak had petered out.

In 1960 Dr Pearce concluded that 'unless it [the fungus] completely changes its present trend of behaviour, it will never bring about the disaster once considered imminent'.[31] Unfortunately for the Atinian Elm, that change happened and enabled the fungus to unlock the genetic defences of that one tree – that one tree with millions and millions of clones.

It was in 1967 that this new strain of Dutch Elm Disease was first recorded in Britain. It spread rapidly, and by 1990 there were very few mature 'English' Atinian Elms left in the country. It has been estimated that 25 million elms were killed in Britain; France meanwhile lost an estimated 97% of its mature elms and Spain around 90%. It was devastating. The disaster that Dr Pearce mentioned had arrived.

The pedant in me must point out, though, that Dutch Elm Disease doesn't actually kill the tree. Instead it effectively coppices it, with the affected stems dying back down to ground level whilst the rootstock of the tree carries on, throwing up new suckers to continue its life. This is why there are still plenty of Atinian Elms growing in the hedgerows around the landscape of the River Taw today – it's just that they are small remnants of once large trees.[32] What actually kills the Elms is us and our actions; millions of trees were grubbed up and burnt in a futile attempt at controlling the spread of the disease. That is what actually killed the Elms, not the fungus.

Dutch Elm Disease certainly changed the British landscape and many of its riverscapes from our perspective, but the fungus would not have had any effect on Otters during the 1960s, 1970s and 1980s when it was spreading unchecked through the Elms of the British countryside. The trees would have been irrelevant to the Otters; true, a few of them may have provided potential holt sites, but their loss would not have been noticed by Otters simply because whilst Dutch Elm Disease was doing its thing, there were very few Otters left in our countryside to notice. They were already going through their own (human-induced) disaster. By the time their population recovered, by the time the Otters regained the rivers, large Elms were nothing but a memory.

The past hundred years have seen many changes in the supporting cast with which the Otter shares the riverine stage. The majority of those changes will have had little bearing on the Otter and its life, with the Beaver being the greatest exception, but that animal's return is nothing but beneficial to the Otter. Alterations to this cast are happening all the time and will continue to occur as our climate steadily changes. Dragonflies are a good example, with several species considered to be rare and occasional visitors from further south in Europe now breeding here. However, there is one supporting cast member that I have not mentioned, and that is the most important one of all. Humans. How we act on the riverine stage has changed in the last century; our behaviour has changed, our numbers have changed, and those changes will inevitably have had an effect on the Otter.

CHAPTER 6

A Disturbing Enemy?

The interlocking leaves of the canopy shut out the bright blue sky above the wooded valley. Towering oaks underscored with a mixture of hawthorn and holly give way to a tangled, collapsed world of willow alongside the fast flow of the Taw, the latter tree species, happy in the wetter soils, forming what seems like an impenetrable barrier. But this barrier of intermeshed branches is not as impassable as it first seems; alongside the flow of the water is a broad muddy path, dotted with puddles and surfaced with innumerable imprints of human footwear and the criss-crossed pattern of mountain bike tyre tracks.

In Williamson's book, the only humans that Tarka encountered on his fictional journey down the River Taw from its source to the sea were ones who were trying to kill him or witness him being killed. If an actual Otter was today making that same journey from the waterlogged peat of North Dartmoor to the confluence of the Taw and Torridge in Bideford Bay, it is highly unlikely that they would encounter anybody intentionally seeking to harm them. But what is certain is that they would encounter humans, for today we like to spend time by, on and even in rivers.

An Otter that was living on the Taw in the 1920s would likely have very rarely encountered a human, even ones that were intent on killing it. Williamson's book may contain vivid descriptions of hunt days, the noises and colours of the followers, the baying of the hounds as they were encouraged to splash through the waters, the sounds and vibrations of steel-shod Otter poles being struck against the riverbed as the hunt followers formed a stickle to block the Otter's escape – but those actual hunt days would have been few and far between for an individual Otter. A hunt might not revisit a stretch of river until two years or more after their last visit. A hunt day would have been an incredibly stressful experience for an Otter. It may have eluded the hounds, may have evaded them altogether, but the stress caused to the individual animal would have been extreme, and that stress would have undoubtedly affected it for far longer than the actual hunt day itself. A major part of that stress would probably have been caused by the fact that the Otter in question had never experienced humans before; everything about the hunt day would have been new for it.

Otters today share their riverine habitats with humans, far more so than they have done in the past, and with that increased sharing comes the potential for increased stress. Our intentions when we visit waterways may be completely opposed to those of the Otter hunters in the 1920s, but for the Otters themselves they may be just as stressful. The noises and colours of the general public, the barking of pet dogs, often several at a time, as they are encouraged to splash through the waters, the sounds and vibrations of people either throwing stones or themselves into the water – all will have an impact on any Otters that are hiding away in a nearby holt. Otters today are constantly disturbed by human activities. The disturbance is far more regular than an occasional hunt day, and that might mitigate the stress the animal feels, but any disturbance to a wild animal is going to cause it stress. Any stress will in turn have an adverse effect on that animal's well-being. Otter hunting is rightly viewed as being abhorrent by the vast majority of people, but those same people often do not think about the likely impact of their own behaviour on Otters and other wildlife.

Most animal species have declined, some drastically, in the UK since 1927, but two species have not, and in fact have done the opposite, increasing exponentially. In 1927 there were just over 39 million humans in the UK, whilst today there are over 69 million of us. Whether you express that as a 77% increase or simply as an extra 30 million people, it is a huge rise in the population of a species that has a massive impact on all other species.

But it is not just our numbers that have changed; our lives have changed too. In 1927 the average working week in Britain was between 48 and 50 hours, depending on whether it was winter or summer – less daylight presumably accounting for the lower number of hours worked in the winter. At the end of 2024, the average working week in Britain was 36.8 hours, a difference (in summer) of 13.2 hours a week. In 1927 paid holiday from work was something many people aspired to, but it took another eleven years for it to become reality for them when, in 1938, the Holidays with Pay Act gave the working person one week's paid leave a year. Today, for a full-time employee working five days a week, the paid holiday entitlement is 28 days or 5.6 weeks a year. Not only are there a lot more of us in Britain today compared with a hundred years ago, but we also have a lot more leisure time too.

We fill our leisure time with all sorts of things, whether that is staring at various screens, reading a book like this, being part of a club or activity, or visiting places. And one of the places we like to visit above all else is the place that we call the countryside. It is impossible to compare between 1927 and today when it comes to visits to the countryside. There are no statistics for the 1920s for a start, and our entire country's demographics have also changed, from being a largely rural-based population in the 1920s to now very much a largely urban one. The county of Devon, home of the Taw and the fictional Tarka and a county whose occupants are often thought of as being very rural, is a good example of this: the county has nearly half of its population living in the two cities of Exeter and Plymouth and the urban area of Torbay.[1]

In general, today we work a day and a half less a week, we have much more paid holiday leave, and we have far greater access to transport to enable us to travel from where we live to countryside areas. Whilst most of us today may live in urban areas, we are very much drawn to the countryside when we have free time; in 2023 we made 219 million day visits to it. Bearing in mind as well that there are now 30 million more of us than in 1927, it is safe to say that the countryside today is far busier in terms of people than it was a hundred years ago. Rivers are not just found in the countryside, of course, but the majority of riverine miles in Britain are. Otters today will be much more habituated to our presence in the countryside than they were in Tarka's day. Our increased presence in the countryside will inevitably cause an increased level of disturbance to Otters and other wildlife, but it is perhaps what we often bring with us that causes the most disturbance.

The second species that has increased its British population so dramatically since 1927 is one explicitly linked to humans: the domestic dog. Even though dog licences had been introduced back in Victorian times,[2] there are unfortunately no accurate figures for the domestic dog population of Britain in 1927. However, by extrapolating out data from a variety of sources it is possible to come up with an estimate of between 600,000 and 800,000 domestic dogs for the time. Today, the number of domestic dogs in Britain is a staggering 13.5 million. In percentage terms, in the last hundred years, Britain's domestic dog population has risen by around 1,800%, a population increase that makes our own look insignificant. It is often said that we are a nation of dog lovers, which also makes us a nation of dog walkers, and the countryside and its rivers are popular spots for walking the dog.

The domestic dog that many of us share our lives with is a carnivorous predator. Genetically speaking our pet dogs are virtual wolves, sharing over 99% of their DNA with the species they were originally domesticated from over 14,000 years ago.[3] Wolves, as we have seen, have been absent from Britain for hundreds of years, and will have been absent from large parts of it for over a thousand years, but their genetic memory, innate in other species of wildlife, lives on. I spent 20 years as a forest ranger in various parts of southern Britain. In that time I witnessed countless examples of the disturbance caused to wildlife by domestic dogs, disturbance that in some cases could only be described as an innate response rather than a pragmatic one.

Red Deer are our largest land mammal, with mature stags weighing in at 200 kilograms and standing as tall as 1.2 metres at the shoulder – they are big animals. Large they may be, but I have seen these impressive-looking beasts panicked into flight by small dogs that I could have picked up with one hand. On every occasion I have seen this, the deer had not seen the dog; it was not the actual sight of a domestic dog that caused the many times larger deer to flee, but the sound and most probably the scent of it. Somewhere, buried deep in the make-up of the deer, is an innate response to that sound and smell, a response to avoid the cause. A Red Deer, whether stag or hind, is more than capable of defending itself against a small dog. They are very adept kickers and more than capable of aiming a sharp, powerful hoof at a foe.[4] But Red Deer, like our other wildlife, evolved alongside the Wolf and the need to distance themselves from Wolves evolved with them. The response I have seen on far too many occasions wasn't necessarily a response to the actual dog, but a response to a historical threat.

It is not popular to say so, but domestic dogs cause massive disturbance to our wildlife; their presence alone is enough to raise stress levels and alter behaviour in our wild species, be they birds or mammals. This is easy enough for us to see if we spend time anywhere where people let their dogs off the lead, and one of the popular places for people to do this is alongside watercourses. The dogs are often encouraged to enter the water, from where they can explore parts of the river and its banks that to which they would otherwise have no access, sites that could well be used by Otters as couches or even holts. Today, there are more domestic dogs in our rivers and waterways than there ever were Otterhounds.

We can see this disturbance ourselves, but the fact that they are causing stress to wildlife is also backed up by numerous research papers and articles.[5] Yet many people still do not associate their dog walking with causing wildlife species harm. Indeed, many people who work in the 'wildlife and conservation industry' often do so accompanied by their dog. In the last hundred years our own population and the population of our domestic dogs have increased substantially, and at the same time so has the amount of leisure time we now have available to visit and enjoy areas such as rivers. As a nation we simply love messing about on the river.[6] The disturbance that rivers and their wildlife now experience compared to a century ago has as a result also increased substantially, but it is something the majority of us seem unable, or perhaps unwilling, to notice.[7] But, as we shall see in a later chapter, this disturbance isn't the only thing that a dog splashing in a river brings.

A night shift?

Otters are famously difficult to see, particularly in England, but is that down to how we behave rather than just their nature? Otters are generally thought of as being nocturnal, in that they are typically active between sunset and sunrise and typically inactive between sunrise and sunset. But this behaviour does not occur everywhere in Britain, especially in coastal areas in western Scotland and the Hebrides where Otters can be seen frequently during daylight. There are two hypotheses here. The first is that in the areas where they are seen during daylight there is significantly less disturbance from humans (and dogs).

This is a tantalising theory, and certainly sounds plausible, but it is also very difficult to back up with hard facts. The second hypothesis is that the saltwater fish that the coastal Otters are feeding on behave differently to freshwater fish and as a result, they are easier for the Otters to locate and catch during the daytime. Many coastal fish are inactive during the day, hiding under stones or in areas of dense seaweed, sites and behaviour that an Otter can readily exploit; after all, an inactive fish is far easier to catch than an active one. However, one of the Otter's favourite prey items in rivers is the Eel, yet these are very active at night and largely inactive during the daytime. It therefore should be easier for the Otter to hunt them during daylight hours.

With sea water comes tides, and these are also a factor that should be considered. The constant ebb and flow of the coastal waters greatly alters the feeding environment for many species, be they wading birds or Otters. Could certain points in the tidal cycle make hunting fish easier (and conversely, harder) for Otters? If so, it could explain the difference in activity patterns seen. Coastal Otters also eat a lot of crabs, a prey item whose availability is heavily influenced by the state of the tide; they are often cut off and trapped in rock pools during low tide and therefore, you would suppose, become easier to locate and catch.

There is no easy answer to why Otters are nocturnal, but I think it is well worth exploring the human-induced disturbance factor a bit further. I spend a lot of time in central Spain, specifically in the region of Extremadura. I have lived there permanently in the past and still spend parts of the year there. I have never seen a wild Otter in the daytime in England, despite spending lots of time by waterways where, as their spraints attest, Otters are present. Yet in Extremadura I have seen Otters in broad daylight on many occasions. There will be differences in various aspects between England and Extremadura – different species of fish, for example, and certainly different climatic conditions – but the one big difference immediately noticeable to me is the amount of human disturbance that occurs along the rivers in the areas.

I will spend hours at a time sat in my favourite river valley in Extremadura; here I am able to engage in one of my preferred pastimes, vulture gazing. As well as these gargantuan avian wonders there are also numerous other bird species to see, everything from mighty Spanish Imperial Eagles to miniature Zitting Cisticolas. It is a place where I can literally lose time, often for several hours. It is also a place where I can, and do, watch Otters in the daytime. But this site is not some far-off remote valley that I have had to trek for hours to reach. It is just four miles from the town where I have a house and is right alongside the main road that leads from it. Beside the bridge that forms part of this road is a large parking area complete with a picnic site. It is in a remarkably scenic spot as well as being in a very accessible location, but it is also somewhere I very rarely see another person. And if I do, they will typically be sat in their vehicle rather than walking about.

The river itself is very easy to access from the car parking area – a broad stony track leads down to it (the large stones at its side next to the river often feature Otter spraints) – but to see someone else down by the river is a real exception, and those that I have spotted have been either a farmer checking on his wandering cattle or a lone fisherman. I have never seen groups of people as I would expect at a similar site in England, and I have never, ever seen a domestic dog. Dog ownership in that area of Spain is very low compared to England. It is unusual to see one being exercised in the town, and in the countryside around the town it is something you just do not encounter.

It is tempting to conclude that the reason I see Otters relatively frequently during the daytime in Extremadura, but not at all in England, is that there are a lot fewer people visiting the rivers in the former, and the few that I do see are not accompanied by dogs. That is perhaps a bit too simplistic – there will always be a mixture of factors that contribute to an animal's behaviour – but our presence, and the presence of the dogs we bring with us, will have an impact on that behaviour.

I wonder if it is another case of the shifting baseline syndrome we talked about earlier. Many European mammals that live in relatively close proximity to humans are largely nocturnal in their activity. We are ourselves a diurnal predator, and although we today normally contain our predatory hunting activities to supermarket aisles and online sites, for the vast majority of our existence we have hunted what we have found in the environment around us and have done so during daylight hours. Our behaviour over the many millennia that we and our predecessors have been present in Europe will have undoubtedly shaped the behaviour of the wildlife around us. And that begs the question: are these animals nocturnal because of us? Today we accept it as normal that they are nocturnal rather than diurnal, but was it always like that? Did that baseline of behaviour shift at some point? Personally, I believe it to be very likely, but whenever it happened it would have been a very long time ago. You may disagree with me, wildlife researchers may disagree with me, but we should never underestimate the impact we humans have had, and still have, on all the other species with which we share this planet. Humans, we are now realising, have been the key factor in the extinction of many of the megafauna that once roamed the continents. Why can't we be the factor that influences the behaviour of the species that are left?

However, humans and Otters can and do coexist. We have already seen that Otters can now be found and seen on a regular basis in riverscapes through built-up towns and cities; here they appear to have become habituated to our busy, noisy and disruptive way of living. They cope with the disturbance we produce in these environments, filling the available territorial niche despite us. But what is key here is that it is the animal that has chosen to cope with everything urban living brings with it. It is a remarkable thing to happen, if you think about it.

As our towns and cities increased in size and noise and human population in the latter half of the twentieth century, Otters were severely reduced or most probably absent from the urban (and countryside) stretches of rivers that ran through them. But as they recovered in England at the end of that century and the beginning of this one, as they slowly regained the lost ground, filling territories one by one, slowly expanding river system by river system, county by county, they would have encountered something their ancestors would never have experienced: modern-day urban living and all that it brings. Nature is adaptable, Otters are adaptable, but we must not forget that it takes time for animals to habituate themselves to areas where we thrive in large numbers. Just because some of them do, it does not mean that all of them are comfortable with the disturbance we cause.

A Peregrine Falcon nesting on an exposed crag on a remote moorland may take to the skies alarm-calling at the sight of a lone human figure 500 metres away, leaving its precious eggs at risk. Conversely, a Peregrine Falcon nesting on the ledge of a building in a busy city may snooze peacefully sitting on its eggs whilst below it, thousands of people move in all directions, vehicles chunter along and sound their horns, pneumatic drills rip up concrete, a police car and a screaming siren flash by and yet the Peregrine, a bird that can be very sensitive to disturbance when nesting, doesn't bat an eyelid at the general acoustic bedlam that forms our urban soundscape.

Likewise, a female Otter used to urban living may be happy breeding in the heart of a city, but one that has spent her life on a quiet stretch of river might give up on a breeding attempt if we decide to increase the number of people around that particular stretch of water by building a new surfaced path to encourage human (and dog) visitors. We must be careful not to fall into the trap of thinking that all Otters will be happy at living with increased human-caused disturbance, just because some Otters are now making their home in our cities.

Wherever Otters live in Britain today, be they in cities or rural areas, they are likely to come across one part of the human infrastructure that we take for granted. For them though this feature is incomprehensible, for they do not understand its significance; and that lack of understanding can have fatal consequences.

Why Did the Otter Cross the Road?

The dark waters quicken noticeably as they approach the bridge. The surface of the river loses its irregularity and takes on a silken aspect as it pours itself through the narrow tunnel-like structure, taking a brief journey through a dusky darkness before emerging into the light again on the other side. The silken smoothness breaks down into eddies and turbulence as the River Taw widens itself out once more. The river has negotiated its first road.

The Taw flows for six miles before it comes to its first road bridge; that is a long distance for an English river. England has a vast road network that criss-crosses the land and its rivers, and the county of Devon has the largest road network in the country,[1] yet for six miles this Devonian river flows without encountering tarmac, without having to negotiate a road crossing. Rivers of course negotiate these structures with ease, their pathway through them engineered by us humans, but the river's wildlife does not necessarily navigate them so simply, and for Otters they can at times be positively dangerous places.

When Tarka made his fictional journey down the Taw, he would have passed by this bridge, or at least an earlier version of it. The road that travels across the structure has changed many times in the last hundred years; at one stage it was the major trunk road taking flocks of tourists on to Okehampton and the far south-west beyond. The volume of traffic that travels across it has fluctuated over the years, much like the flow of the river beneath it. It is tempting to think that road traffic is a modern-day phenomenon and that a century ago the dangers this traffic now poses to us and wildlife were pretty much non-existent. But actually that is not the case at all.

Thirty-one years before Tarka made his fictional journey, Britain had its first motoring fatality. It was on 17 August 1896 that a Mrs Bridget Driscoll of Croydon was knocked down and killed by a vehicle at Crystal Palace in the south-west of London. Whilst that unfortunate woman's name has gone down in history, the first wildlife casualty on our roads was never recorded.

By 1927 when the book was first published, the risk of being knocked down by a motor vehicle would have been a very real one for an Otter, as well as for many other wildlife species and us too. In that year there were already a staggering 1,209,928 motor vehicles registered for public road use in Britain; add in motorcycles and that number rises to 1,900,603 motor vehicles on British roads a hundred years ago.[2] Traffic was already using the first road bridge that Tarka would have encountered on his journey down the Taw. Motor vehicles were thus ever-present and in surprising numbers; they themselves are not a new modern-day threat, but their volume is. By the end of 2023 the RAC reported that there were now 41,200,000 motor vehicles registered for use on British roads.

In theory this means that today an Otter is over 20 times more likely to be hit by a vehicle than it was a hundred years ago. Whatever the merits of that theory and my mathematics behind it, one thing is certain: Otters, and many other wildlife species, get hit and killed on our roads on an all too frequent basis. In one ten-year study, carried out by The Road Lab of Cardiff University, which ran between the 2013 and 2022, a total of 1,838 Otters were killed on our roads, and those are just the ones we know about.[3] Most wildlife road casualties are not recorded – the unfortunate animal is either damaged beyond recognition before it is noticed by someone who wants to record it or, as often happens, it moves away from the road to die of its injuries, out of sight and out of mind. One researcher, talking about all species of animals that are regularly killed on our roads, suggested that they would be surprised if they capture 1% of all roadkill that occurs in Britain.

The Otters of the Taw and its tributaries are part of those roadkill statistics. The Otter Project also based at Cardiff University has produced a map showing the locations of reported Otter road casualties across the country, each sad demise marked by a red dot.[4] The first red-dot locations as you travel downstream from the roadless high land of Dartmoor occur on the first roads the river encounters on its journey. As I continue to scroll through the zoomed-in map, following the Taw river system northwards, it does not take me very long to find another red dot, then another, this time only about a half a mile from where I write this.

As I scroll, so the red dots keep appearing. I soon find a cluster, three records on a three-quarter-mile stretch of the main A377 road as it snakes alongside the sinuous meanderings of the Taw. Before I get to the town of Barnstaple, before I get to the point where the river becomes an estuary, I count 15 red dots – 15 dead Otters. All of these are along a stretch of river system of around 25 miles in length, but these are not the only Otters of the Taw that would have died; the map has a disclaimer above it clearly stating that it is not a comprehensive list of mortalities. The 15 Otters, the 15 red dots, are just the ones reported to the project.

We will never know how many Otters on the Taw were dying as road casualties during the era that Tarka was written, but one would suspect it was a much lower number than today. As we have seen, there were fewer vehicles on the road (although still a surprising amount) and of course those that were travelling the Devon highways in the 1920s, such as the

A377 Exeter to Barnstaple road, would have been doing so at lesser speeds. The risk to Otters from roads was present a century ago, but today that risk is greater than ever before, especially elsewhere in the country where Otters have only recent made a return. When they were last present in some of these areas in the mid-1950s, there would have been fewer roads as well as fewer cars.

The figure of 1,838 known Otter road casualties from the study by The Road Lab equates roughly to one Otter being reported as killed on our roads every two days, but Otter road casualties do not occur evenly throughout the year; the seasons and their weather have a large influence on when they take place. Of all the species recorded in the ten-year Road Lab study, the Otter was the only mammal that demonstrated a winter peak in reported casualty numbers. Otters are much more at risk of becoming roadkill in the winter months than they are at other times of the year. The answer to the question 'Why did the Otter cross the road?' is quite simple: the weather. In Britain our unnaturally linear rivers typically flow faster and higher during the winter months than in the summer months, due to the increased rainfall we receive in the winter. Otters are very skilful swimmers, but high levels of water being forced through restricted structures like bridges can make them very dangerous places for even the most experienced of swimmers.

Rivers in spate are potentially dangerous places for Otters; they are also places where feeding opportunities are hard to come by, and where holts can become inaccessible. Natural rivers are a rarity in Britain. Instead of dynamic wetland habitats that can slow the flow down, rivers are generally artificially straight channels through which excessive water flows fast. As we have already seen, the presence of Beavers can help mitigate this, as they can bring back the dynamism that brings slower flow, but for the majority of Otters in Britain, Beavers are not (yet) present to do so. When rivers are in spate, territory-holding Otters can find themselves forced to move to somewhere they can feed, somewhere they can rest up safely, somewhere out of the dangerously fast-flowing waters. It is instinctive for an Otter to avoid danger, to steer clear of risky, unpredictable waters, but, sadly, in their attempt at avoiding that danger, they can often unwittingly place themselves in yet more peril. Otters can and do cross dry land, but they prefer to use watercourses to navigate their way. In times of high rainfall, even if the watercourse itself is swimmable, they will inevitably come across a section of river that is impassable to them. Invariably this is the section of river that passes through a human-made constriction, a structure built across the river that forces the water underneath it – in other words, a bridge. Rather than risking swimming through these speedy, powerful waters, Otters tend to opt to go around the obstacle. To do this they leave the river and take to the land; they are forced upwards out of the water, they are forced to cross the road.

It will come as no surprise that around 65% of all recorded Otter roadkill casualties occur within 100 metres of a watercourse, and 34% of these occur at bridges. But bridges are not the most dangerous constructions; 44% of Otters found dead on the road close to watercourses die at culverts. Culverts are structures that channel water through or past an obstacle, and are typically embedded into the ground, the only route through being the channel of the culvert itself. They are effectively miniature, tunnel-like bridges, but their small size means that in times of high water flow they can be completely filled up with fast-flowing water. When they are dry, or when there is little flow travelling through them, they are the perfect routes for Otters to take, for they are in effect mini subways beneath a road. But when full of rapid water they are a danger that the Otter instinctively looks to avoid.

Increased rainfall leads to increased Otter road casualties. Winter is generally the wettest season in Britain and the number of road casualties at this time increases accordingly. Data recorded by The Road Lab team at Cardiff University shows this clearly. The number of Otter road casualties recorded in May, June, July and August is very low, as are typically the rainfall amounts and the corresponding river levels. The number of casualties starts to increase in September and October as rainfall and river levels rise once more, whilst November and December are higher again; but it is the months of January, February and March that have the highest number of Otter roadkills, with the shortest month of the year, February, showing the greatest of all. It is in these months that the rivers in Britain run at their highest, and it is these months that are therefore the most dangerous when it comes to Otters crossing roads.

But in recent years, due to climate change, our weather in Britain has become more unsettled at other times of the year. We are increasingly getting periods of very heavy rain during the summer and spring months too, rainfall that can be so sudden and so intense that it can cause flash flooding as watercourses rapidly fill and struggle to cope with the increased flow. As an example, the river that runs by the edge of my village before it meets the Taw a few miles later will rapidly fill up and become a torrent under the main road bridge following periods of heavy rain. In the past this would have invariably only been in the winter months, but increasingly it is happening at all times of the year. It is highly likely that the changes we are seeing in rainfall patterns, their amounts and their intensity in the spring and summer will lead to more Otters being hit on the roads during these periods, as the animals seek to avoid the dangerous conditions this unseasonal rainfall creates through structures like bridges and culverts.

There is something that we can do, though, to try to make these structures safer for Otters during times of fast and high water: the provision of a route through the structure that is above the high-water line. These routes are given various names by the companies that market them, with 'wildlife shelf' being commonly used; it is a good name, as it describes the new route well. A simple ledge or 'shelf' is installed along the inside of the structure providing an elevated route through it – these can be retrofitted or, better still, they can be incorporated into the design of the structure in the first place. When one of these shelves is combined with fencing, which helps funnel the Otters (and other wildlife) to the shelf route, they can provide effective and safe passage, but they are only suitable for some structures, typically larger bridges. Culverts provide much more of a challenge, for they are by their nature small structures with restricted space. Short of ripping up and replacing them with new and bigger structures, there is very little that can be done to mitigate the risk they pose to Otters when water levels are high and fast. But where it is feasible to install such devices, they should at the very least be considered.

Otters have only relatively recently recolonised West Sussex in the south of England following the pesticide-induced population collapse. Since their disappearance from the county in the 1950s and 1960s, the road network here has grown, as has the volume and speed of the traffic that uses it. It is thought that their return to the county's rivers has been hampered by this new and busy road network, and that Otter casualties on these roads could be an important limiting factor in their continued recovery. Staff at West Sussex County Council recently identified an area where breeding Otters had returned, a joyous moment for anyone, but they also realised that a nearby road bridge for a busy main road on the river system posed a potential threat to these animals, and to any future dispersing progeny of theirs, if the water was running high. Their solution was to install a wildlife ledge on the bridge, providing safe passage to the Otters when required. This is a great example of proactive conservation – no Otters had been run over on the busy road, but the potential for it to happen was certainly there and was identified in advance. It has taken a long time for Otters to return to this part of England, and the prospect of them disappearing once more was a risk that I am pleased to say the council were unwilling to take.[5]

Road signs can also be used to warn motorists of the potential of Otters crossing roads. They take the shape of the familiar red-fringed warning triangle, but instead of the silhouette of a cow or a deer within, they instead feature the unfamiliar outline of an Otter. These have been erected at notorious Otter casualty blackspots in various parts of the country. The Taw itself has a couple, installed on the busy A361 road that flanks the Taw estuary after it has flowed through the town of Barnstaple. These were paid for by the UK Wild Otter Trust (whom we will speak more of later), who had identified the site as being particularly hazardous for Otters, with around 16 being reported by the Trust as killed on this short stretch of road in recent years. It is hoped that the signs will make drivers think about their speed, make them more aware that wildlife such as Otters could be encountered.

There are mitigation methods that we can employ to reduce the risk to Otters from roads; various adaptations to bridges and structures can be and have been employed across the country. Road signs can be installed to raise awareness and perhaps even encourage lower speed. But the best mitigation method we can use is to allow our river systems to act more naturally, rather than being canalised channels designed to get rid of water quickly, something we are now realising leads to rapid rises in water levels further downstream in periods of intense and prolonged rainfall. We need to allow our rivers, wherever possible, to be natural dynamic systems, reducing the risk of not only flooding but also wildlife road casualties. The return of the Beaver, as we have seen, can help this process, but in reality we need a paradigm shift in our own thinking for it to truly happen.

There is space for this change to take place, but it is a restricted space and is growing smaller every year due to increased development, development that often takes places on what were once called floodplains. We cannot, and nor should we, expect farmers to allow their land to be continually flooded either – we need food, after all. But we need a national plan for our watercourses that not only makes them better places for our Otters and wildlife, but also for us. Our rivers need to be more robust systems; our climate is changing, we are getting increased spells of wetter weather with more intense rainfall. We need a plan in place, and this is a topic that we shall return to in the final chapter. Whether it will happen, though, remains to be seen.

In the short term, however, mitigation methods such as providing warning signs and ledges underneath bridge structures are a good thing, but we must not be fooled into thinking they are the solution.

When there are problems between humans and wildlife, we tend to refer to them as 'conflicts', a word that has many definitions with, according to the *Oxford English Dictionary*, its main usage today describing interests that are incompatible. But the needs of Otters (and wildlife in general) should not be incompatible with our own needs as humans; there needs to be coexistence. The conflicts mentioned in this and the previous chapter, the conflict between human (and dog) disturbance and Otters, as well as the problems that roads bring to so many wildlife species, are down to a lack of consideration of the needs of Otters and other wildlife in the way we live our lives. We need to think about everything we do in a light that also considers the needs of all the other species with which we share this planet. Only then can we have true coexistence.

We might not think of driving our car as being part of a potential conflict with Otters (or other animals), nor do most people consider their presence in wildlife habitats as being so either. But there is one thing we do, and millions of us like to do it in Britain, that is often seen as a conflict between humans and Otters, and that is fishing.

CHAPTER 8

Fishing for Answers?

The velvety sheen of a Moorhen glides through the water, creating a mini bow wave that ripples out across the large, tree-fringed lake. A Robin sings its slightly melancholic but thoroughly beautiful song from an unseen perch, whilst a Blue Tit hangs acrobatically off the end of the flimsiest twigs as it scours the leaves for sustenance. Suddenly, a clear piping call cuts through the tranquillity of the moment and a flash of electric blue zips low across the water, an avian tracer bullet that alights on a willow branch on the far side of the lake. A Kingfisher. It sits motionless on the branch, taking advantage of its position over the water to stare down its beak at the watery world below. The bird has found its fishing site; the king of anglers is in position. But it is not the only one.

The lake is actually the flooded remains of an old gravel pit, the spent industrial workings now full of water rather than the geological deposits that generated a business opportunity here in the middle of the last century. The gravel may be gone, but there is still a business opportunity here – it's just the nature of it that has changed. This is a commercially run carp fishery and its assets are highly valued.

Fishing rods emanate out from the lake edge, long straight poles evenly spread around the perimeter of the lake, each one belonging to an individual who is sitting quietly, pitting their wits against the large carp that dwell beneath the surface. This is carp fishing and it is big business.

Carp is a generic name given to a large group of closely related fish; of these the Common Carp is probably the best known of the various species. It is not native to Britain, originating from eastern Europe, but its large size combined with the fact that it can be easily kept in large ponds means the fish has long been spread to new areas by humans, so that it can be used as a reliable and readily available food source.

In Europe, the Romans seemed to have started the practice of farming carp for food, creating ponds that were then stocked with the fish at higher than natural densities.[1] Here the fish could be easily caught and removed whenever one was required for food; the ponds were in effect a living parlour cupboard. The Roman Empire crumbled in the fifth century, but farming carp for ponds did not disappear with it. The newly formed monasteries of Europe kept the practice going, developing it and spreading it further, the fish fitting very well into their Catholic diet as a land meat substitute during times of religious fasting.

No one knows for sure when carp were introduced to Britain. There is much debate on the subject; some say the thirteenth century, some say the fourteenth and some say the fifteenth, their actual arrival lost in the murky, deep pond waters of time. What is known is that by the year 1600 carp ponds were well established in Britain and the fish had become widely naturalised across the country in neighbouring rivers. In 1653 Izaak Walton wrote and published *The Compleat Angler*, a discourse on fish and fishing in Britain; in the book he calls the carp the queen of the rivers, stating that it is a stately and good fish.[2]

Carp are chiefly fished for in this country as a sport fish rather than for eating. Commercial carp lakes stipulate that all fish caught must be returned to the water, and are not allowed to be taken. The attraction for British carp anglers lies in the catching of the fish rather than its eating. And that attraction is great: carp fishing in Britain took off in this country in the 1950s, before really exploding in popularity from the 1990s onwards. In an Environment Agency report covering 2015, carp fishing made up a third of all 'fishing days' in England with a total of 7,440,000 days dedicated to it in that year.[3] That's a lot of days, and it's also a lot of money – the same report valued the contribution of angling to the country as being over a billion pounds.

Carp fishing really is big business in Britain, and with the growth in its popularity has come a growth in the number of businesses running commercial carp lakes. These started to be opened in the 1950s and have continued to open ever since, with many starting life in the 1980s and the early 1990s as carp fishing began to really boom. These dates are significant.

The conflict between humans and other species that catch fish is long ingrained. The Tudor Vermin Acts we examined in Chapter 2 amply demonstrate our attitude to anything else that dares to catch a fish (and usually succeeds far more easily than us). Putting a bounty on the birding beauty that is a Kingfisher says more about our intolerance of other fish eaters than any words that I can write. Today, whether the fish that we are trying to catch are being sought for food or for sport, competition for them isn't always appreciated. I spoke to one fisherman that I know, someone who happily talks about his delight at seeing an Otter, who told me very openly that he hates it when he is at his local fishery and sees someone else on the other side of the lake land a fish. To him it seems unfair that someone else has caught it, and there are no congratulatory thoughts on his part – just envy and, as he himself admits, a little bit of anger. It is no wonder then that Otters are often perceived by people that fish as being a problem, even a pest.

Otters eat fish, and carp are fish. Otters like to catch slow fish, and carp are slow fish, particularly when water temperatures are low, as they are in many deep ponds for most of the year. Ergo, Otters eat carp. A large pond full of large ponderous carp must be absolute nirvana for an Otter, an all-they-can-eat buffet they can return to day after day, catching fish at ease in the still waters of the fishery. It is not the Otter's fault; the animal is not looking for conflict, but is simply exploiting an abundant and readily available food source. But it does of course bring it into conflict with the carp fishing business.

As carp fishing as a commercial enterprise started to develop in the 1950s, so Otters were beginning their catastrophic decline, one they only really began to recover from in the mid-1980s and into the 1990s. Even then, that recovery was initially very slow. As we explored earlier, the fourth National Otter Survey of England in 2002 showed that only around one third of all sites surveyed had positive signs of Otter; many parts of England still lacked them. When the boom in carp fishing happened, when new carp fisheries were being developed in the 1990s, Otters were still at artificially low numbers. They were still absent from whole counties, so the threat of Otter predation to carp fisheries was not really an issue in most parts of the country, and thus was easily overlooked. But as the Otters regained the waterways, as dispersing youngsters sought out new territories for themselves, they soon discovered these significant food sources, and Otter predation rapidly became a problem that could not be ignored for many fisheries and their anglers.

A quick internet search on Otters and carp will instantly bring up articles and opinions on the topic of conflict between Otters, carp fisheries and their anglers. The vast majority of written material squarely puts the blame on the wild animal that is simply behaving in an instinctive way – the Otter is portrayed as being the problem. And it is a problem, even if it is actually not the Otter's fault. It has become an increasing issue in recent years as the Otters make their comeback complete. I am a conservationist, nature is my primary driver in life, and seeing Otters is an amazing experience and one that I will never grow tired of. But I am also a realist, and Otters will cause problems for carp fisheries, I cannot pretend otherwise. No predator is going to turn down the opportunity of an easy meal. But we should not forget that the problem is caused by us creating vulnerable ponds and lakes full of overstocked Otter food.

One hundred years ago the solution would have been both simple and brutal, but we do not live in the 1920s any more. The wanton killing of Otters portrayed in *Tarka* as normal and everyday is not something that would be tolerated now; indeed it is something that is rightly illegal. Coexistence is what is needed, and for coexistence to work there needs to be an acceptance that, without any action, a recovering Otter population will have impacts on the coarse fishing industry, and costly impacts at that.

Not sitting on the fence

In researching this part of the book, in looking at many online carp fishing resources, it was pleasing to see that many in the carp fishing press and community know just how highly Otters are thought of by the wider general public and recognise their recovery from the population crash last century as a good thing, despite the issues it can bring. One influential angling columnist writes 'we must accept that they [Otters] are here to stay', before going on to say that carp fisheries must take action to protect their fish in much the same way that a chicken keeper will shut their chickens in at night to protect them from predation.[4] In short, the columnist was saying that fisheries need to be fenced.

Otters, like many fellow mustelids, can be very tenacious and resourceful when it comes to overcoming obstacles. A quickly erected stock net fence is not going to do the job. For the fence to work it needs to be designed to be Otter-proof, and this costs money, but then so do the carp that the fence will protect. A carp weighing in at around nine kilograms, or twenty pounds in fishing terminology, is worth between £500 and £600, whilst one weighing in at thirty pounds (13.6 kg) could be worth £2,500. Carp fishery lakes, by their very nature, contain a lot of carp; they are deliberately stocked with large fish at far higher than natural densities. Therefore, to create a good carp lake that is popular with carp fishers, who are of course attracted by the potential of landing a large fish, the owners of the business must spend a considerable sum of money on the fish with which they stock the lake. Considering the significant initial outlay on stock, spending money on fencing seems to me, and the columnist quoted above, a very good and very sensible investment. Fencing Otters out of a fishery does cost money, but for the price of three or four large carp, all the fish in the fishery (and there might be thousands of them) could be protected.

Otter-proof fencing for fisheries is widely available, and there are even specialist companies run by keen anglers that offer advice on the subject as well as an installation service. But there are also plenty of very proactive fisheries that produce their own fencing to help solve and/or prevent any potential conflict with Otters. One such proactive fishery is Sanctuary Lakes in North Devon, lying within the catchment area of the River Torridge, the second river of Henry Williamson's Two Rivers Country. I spoke to John Dawe, the owner of the fishery, about what they have done on their eight-acre site, which comprises five lakes in total, to help remove this potential conflict.

The fishery is stocked with carp and Tench, among other fish species; as they breed it is impossible to know how many fish are in the fishery, but John wondered if there could be as many as a million. Whatever the number, it is a considerable amount and many of them are now valuable fish due to their size – they are the assets of the business. The fishery was created on land that was not close to any rivers, and to begin with Otters were not on the fishery's radar. However, Otters are able to access these sites, reverse-following the flow of water from them via drainage ditches and small streams until they reach what is, to them at least, an abundant food source. It did not take long for the Otters of Williamson's Two Rivers Country to find the site, and once they did their arrival was very quickly noticed. As is often the case when a predator finds itself amongst an abundant and easily catchable prey source, the Otters inflicted a lot of damage to several of the fish, often eating only small parts of them before moving on to another. The bodies of carp and other fish were left strewn along the banks of the lake, an ottery calling card that could not be ignored.

John Dawe is a very practical man. His willingness to get stuck in belies his octogenarian status, and he immediately started looking for solutions. The first action they tried was to install standard electric fencing around each of the individual lakes, but this didn't work for a number of reasons, including that it posed an issue for the people actually fishing the lakes. But perhaps more significantly, the Otters were able to get through it. Enclosing individual lakes requires a lot of fencing and the creation of a lot of access points for the anglers; every access point is a potential weak spot in the defences, so the fewer the access points, the stronger the defence. As a result of their experiences with their first fence, the fishery scrapped the idea of fencing individual lakes and instead opted to install one fence with one access point, around the perimeter of the site, reducing the amount of materials required and minimising the number of weak spots. The new construction was made of chain link fencing, like the wire mesh fences often seen around play parks and so on, where they prove an effective barrier against dogs. But the tenacious Otter is very different from a dog; they can climb. And that is what they did, simply clambering up and over the new barrier.

At the same time John also realised that the low-level trees and scrub surrounding the site provided the Otter with a means of circumnavigating the fence, as they could use the larger overhanging branches as a bridge to cross it. The Otters had rapidly learnt how to defeat the latest barrier to their potential food, but John is not a person who gives up easily. So after a return to the drawing board, he was back with an idea for a new fence which was duly installed. The new fence line ran a few metres in from the boundary, creating a gap between it and the scrub and trees on the perimeter. To reinforce this any branches that were deemed a potential threat were cut back to ensure there was clear air between the new fence line and the scrub for its entire length. The fence consisted of composite posts dug into the ground, with holes then drilled through the posts so that a four-strand electric fence could be run through it. Unlike wooden posts the composite posts do not earth the electric current and with the fence being away from the boundary, overhanging vegetation was not present to cause short circuits either. John and his team then waited to see if their labours were successful.

The new fence works. John tells me proudly that in the two years since the fence was installed they have had no predation issues caused by Otters. Their solution is doing the job and doing it well. Obviously, though, this fence came at a cost, but it was not a cost entirely borne by the fishery. The Angling Improvement Fund (AIF) is a scheme funded by the income generated through the sale of the Environment Agency's rod licences; fisheries can apply to the scheme for help with predators and that is exactly what the fishery at Sanctuary Lakes did. The AIF covered half of the £4,300 bill for the fence, so that the fishery had to make an outlay themselves of just over £2,000. When you consider the value of individual fish within the lakes, this is a relatively small amount. The Environment Agency also uses the AIF to fund the work of fishery management advisors so that people like John can access expert advice free of charge. The AIF and the resolution of people like John Dawe are great examples of how proactive approaches to resolving conflicts can work. The Otters have made a comeback and with that comeback comes potential problems for some sectors, but as Sanctuary Lakes and the AIF have demonstrated, these are problems that can be overcome.[5]

Any business, no matter what kind, that is holding stock of high value should have measures in place to safeguard that stock – it is common sense. Carp fisheries that refuse to invest in fencing are not running their businesses well.

It cannot be emphasised enough, though, that the underlying reason for the potential conflict between Otters and commercial fisheries such as carp lakes is because these fisheries are artificial situations created by us. There is absolutely nothing natural about a carp lake. The fish are basically livestock, kept animals at artificially high densities, and there is no balance in the system to counteract the threat of predation, as there is in natural conditions. Otters do have the potential to cause financial damage to these commercial fisheries, but commercial enterprises also have the ability to protect their stock through investment.

But not all fish ponds are commercial.

The keeping of koi carp for ornamental purposes began in Britain in the early twentieth century. This practice, as opposed to keeping them for food, originated in Japan, where the Amur Carp was selectively bred to create fish with distinct markings and colouration. It soon became very popular and the habit of keeping these fish as a hobby rapidly spread. In Britain in the 1960s and the 1970s it became a bit of a fashion trend to have koi carp in your garden pond. It is a trend that still continues today, with koi carp ponds still a popular garden accessory.

Distinctive varieties of koi carp are still being bred, and unusual examples, or even brand-new varieties, can attract serious interest and money from collectors. Yes, there are collectors, and they can be very serious – one Chinese collector reportedly paid two million US dollars for just one fish in 2018. For the collectors of koi, it is important that their fish are easy to see, easy to be admired and easy to show off to others. This leads to them being kept in relatively small and shallow ponds where they are forced to be close to the surface, and having higher than natural numbers of fish in the pond means that there will always be some of them readily on view. Of course, a collection of large fish in a small shallow pond is not just going to be attractive to avid collectors. It is also going to be attractive to anything that eats fish, including Otters.

Garden ponds in urban and suburban environments have largely remained out of reach of Otters, especially when the animals were at their unnaturally low density in the second half of the last century. But as their population has recovered, and as they have recovered lost ground, they have been increasingly reported in these environments. Ponds in residential situations close to waterways are increasingly likely to be visited by Otters, and if these ponds are stocked with easily caught food, then the Otter is going to keep on visiting. In August 2023, the BBC reported on a story from Cheshire in the north-west of England concerning a hotel that had noticed that its collection of koi carp, kept in a couple of ponds in their Asian sensory garden, was being stolen by an unknown thief.

The hotel said that it had lost fish to the total value of around £100,000 and suspecting that the thief was a human, they installed a CCTV system in the bid to catch them red-handed. The surveillance cameras were successful, recording the culprit taking a fish from one of the ponds. But it was more a case of getting caught red-pawed, for the thief, if that is the right word, was an Otter. The hotel had previously taken some precautions against herons, installing an electric fence to prevent the birds from being able to perch along the edge of the ponds, but they had not considered the fish to be at risk from Otters. As with the first version of the fence at the Sanctuary Lakes fishery, the Otter was easily able to get through the very basic electric fence that had been installed.

The hotel is situated on the edge of a small village; the houses lie on one side of the complex whilst open farmland lies on the other. Separating the hotel grounds from this farmland is a watercourse, a small brook that languidly flows about 100 metres away from where the ponds are located in the sensory garden. This small brook is a very minor tributary of the much bigger River Dee, a major Welsh and English river that is around four kilometres from the hotel. The stocked ponds are therefore quite a distance from what many would consider typical Otter habitat, but this did not stop an Otter from discovering this potential food source. Otters thoroughly explore their territories, using smaller watercourses and even wet ditches to get to know their patch – they are always looking for opportunities, be they for food or for shelter. The Otter in this case had evidently been using and exploring the brook, a journey that led it to locating the hotel's well-stocked koi carp ponds. It is likely that the Otter was able to detect the presence of the fish from some distance away, as the pond's outfall that made its way to the brook would have been laden with fishy signs. These would have been readily detected by the tuned-in Otter and after discovering the bountiful food source, it was always going to keep returning to it.

The hotel's case is a classic example of how Otters are today coming into conflict with owners of koi carp. If you were stood by the sensory ponds in the grounds of the hotel, you would not be able to see the small brook 100 metres away, and you would have no idea that there is a major river just a few kilometres further away. In short, it is not the sort of place you would think it likely for an Otter to visit. But Otters are highly resourceful creatures and with their population having steadily increased over the last few decades, they are increasingly likely to discover such places.

When I read the article about the hotel and the koi, I was struck by the ingenuity of the Otter. I admit that I smiled to myself at how the Otter had successfully managed to discover and then take this food source. But if I were the owner of the hotel, if I had lost property worth £100,000, would I smile at all?

There are numerous other reports of Otters predating koi carp from people's garden ponds, a reversal of the usual story of people's pets killing wildlife. The latter is something that happens constantly and at a huge scale,[6] yet it rarely garners any attention in the same media that readily report the 'shocking' news that an Otter has killed some fish.

If we are going to keep unnaturally high numbers of large fish in tiny shallow ponds with no cover, we are going to have cases of Otters predating those fish; it is not a natural situation, but one of our own creation. If we want to keep fish in this manner without losses, we need to employ mitigation methods to protect the fish, be that mesh or fencing. If we are not prepared to do that, then we must be prepared to take the losses. Domestic koi carp ponds are effectively very small-scale carp fishing lakes, unnatural systems with no inbuilt balances. It is not a predator's fault if they are tempted by our provision of an unprotected and very easy meal. It is not the Otter that is causing the problem, it is us.

Convenient excuses for inconvenient truths

Our river systems might not be pristine natural environments, but they are still natural and they do still feature balances in the system. The recovery of Otters back into our river systems doesn't have a detrimental effect on the wild fish that are found there; predator–prey relationships in natural environments are long evolved, and are completely natural in the most literal sense of the word.

There have been some who claim that Otters are responsible for our falling wild fish numbers, especially such highly prized fish as Salmon and Trout. In the past these voices mainly emanated from estates that sought to make money from selling their fishing rights to Salmon fishers; they would routinely claim that the fish numbers available for their customers to try to catch were being reduced by the presence of Otters, even when those Otters had actually disappeared from the majority of the country's rivers. It has to be said that these were usually the same people that viewed any predator, be that mammal or bird, as something that needed to be controlled, the very people that opposed any wildlife-protecting legislation. Paul Chanin noted in his book of 30 years ago that he had previously listened to an MP stating that he was not going to allow Otters to attack the Salmon in the river on his estate and that he would instruct his gamekeepers to control (i.e. kill) them, even if they were made a protected species. The fact that this then-MP was the owner of a large game estate and was speaking at a time when Otters had been virtually wiped out by the use of agricultural chemicals speaks volumes about the sort of people who espoused (and in some cases still espouse) these views, as well as being highly indicative of their ignorance of reality.

But it is not just large landowners of the past that hold these views; there are some that even today still think that Otters are the cause of the decline in wild fish stocks. In January 2025, a video of an Otter carrying a fish on the frosted banks of the River Wye in Herefordshire was posted on a local newspaper's social media page. The person posting it was someone who was evidently thrilled to have not only seen an Otter, but also had the opportunity to film it. However, some of the comments that were subsequently posted in response to the video were really quite troubling. They demonstrated that there is still prejudice towards these mustelid predators, even if that prejudice is completely misinformed. Comments, often accompanied with the seemingly necessary 'angry' emojis, included many that said it was the Otters that were the reason there were no fish in the Wye any more. Another person stated that Otters were a bad thing to have on a river and that they would wipe out the fish stocks before switching their attention to eggs and birds, implying that they would wipe them out too. Another posted that Otters destroy rivers and entire ecosystems. One person even suggested that Otters should never have been introduced in the first place… Many of the comments were made by people who had avatars implying that they were fishermen.[7]

Needless to say, there is quite a lot to unpack here! But firstly, I have to point out that there were plenty of other comments in response to the ones above, pointing out that Otters and fish had coexisted on the River Wye for millennia without there being any problem. Many of those posts were from people whose avatars also implied that they were fishermen. One of those fishermen pointed out rather succinctly that the Otters were indigenous to the Wye and therefore not introduced, but that some of the fish in the river such as the Barbel, which he himself fished for, were introduced by humans and therefore would not be there naturally.

It was good to see those counter-posts from the angling community. I know many anglers who would be delighted to have the opportunity to see an Otter. One of them told me that surely signs of Otter being present on a river is a good thing from an angler's point of view – if Otters are there, he said, then it means that fish are there too. You cannot fault the logic. Fishers are not against Otters; despite a common perception otherwise, I believe that the majority of river anglers know that Otters are a sign of a healthy river system, one that holds lots of fish. But there are some anglers that clearly are prejudiced towards the Otter, seeing them as a reason for falling fish stocks. This prejudice being directed by misinformation and a total disregard for the realities of predator–prey relationships should not lessen the fact that the prejudice exists – that it still exists a hundred years after the fictionally described prejudice that was so clear in *Tarka*, and then the real prejudice recorded two decades later in *Ring of Bright Water*.

It is very easy to dismiss comments like the one about how Otters should not have been introduced (presumably in the mind of the original poster, they were introduced by tree huggers or the like), but even though we see them as nonsense, they do reflect some people's views and we should recognise that. These views need to be challenged with the facts, facts that put a stop to the build-up of resentment to Otters, a feeling that will grow if left unchecked, especially in today's world of unverified social media and the like. Anglers are passionate about their sport, their pastime, and they want to protect it. Those that seek to blame the Otter for a lack of fish need to be informed about the real reasons fish stocks are in decline, the real reasons our rivers are in a poor state of ecological health. Their anger needs to be redirected to the right sources, because we should never forget that they too passionately care about our river systems and want to protect them. The real reasons fish stocks are in decline are abundantly clear when it comes to the Wye.

The Wye is a beautiful river system, but it is now also a very heavily polluted one. In 2023 the government body Natural England downgraded its status from 'unfavourable recovering' to 'unfavourable declining'. It has been widely reported that industrial-scale chicken farming to supply our supermarkets is a major cause of this downgrading; the chickens are housed in vast numbers, with an estimated 23 million chickens being raised in the Wye catchment annually. Chickens produce a lot of manure and that manure is very high in phosphates and nitrogen, both of which cause problems for river systems when they get inevitably washed into them by rainfall.

Both nitrogen and phosphates cause algal blooms to form. These can be vast in size, blanketing entire stretches of water. These algae blooms quite literally starve the water of oxygen, robbing it of that vital ingredient for life. In February 2025, a month after the video post, a group of Wye valley residents, who are demanding compensation for the damage done to the river, were reported by several mainstream news outlets to be considering widening their existing legal action against the companies behind the chicken farms to also include the local water company Welsh Water. The group of residents stated that whilst the industrial-scale chicken farming is behind the bulk of the pollution in the Wye, sewage discharges, high in phosphorus and bacteria, were another serious contributor.

To blame Otters for falling wild fish stocks, to blame them for the lack of Salmon or Trout for fishers to catch, to blame them for game estates not having enough fish to keep their clients happy, is completely missing the point. Salmon and Trout stocks in Britain are indeed falling, there is no doubt in that, but this is due to our management of our water systems, and has nothing to do with predators. Otters are nothing more than a convenient excuse that avoids an inconvenient truth. The 'Salmonid and Freshwater Fisheries Statistics for 2023' published by the UK government in December 2024 highlights just how bad things are.[8] According to the report, the salmonid stocks of 38 of 42 principal Salmon rivers in England were assessed as being 'at risk' or 'probably at risk'. Only one of those 42 rivers was assessed as being 'not at risk'. In Wales, all 22 principal Salmon rivers were classified as being 'at risk'. And the main cause of all this 'risk'? Pollution, human pollution.

Apart from the pollution that we are putting into the rivers, there are other reasons behind the widespread decline in fish numbers, and these reasons cause problems for Otters as well. The way we manage, and have managed, river habitats is a major issue for several species of fish; turning rivers from natural dynamic systems into fast-flowing linear features removes much of the varied riverine habitat that these species require. For the Salmon and the Trout especially, these features include shallow-lying gravel beds for them to spawn in. These beds are created by the natural flow of the river, the current moving and depositing gravel as the river meanders and slows. But if the river is no more than a large straight drain with fast-flowing water, these gravel beds will not form. And if they do not form, then there is nowhere for the Salmon and Trout to spawn along that stretch of river, and consequently their breeding success in that river system can plummet.

The removal of trees growing along the banks of rivers as part of ironically named 'river improvement schemes' does not just deprive Otters of potential holts, but also dramatically reduces the shading of the water itself, exposing more of it to direct sunlight. Variety is the spice of life, as we are often told; variety of water temperatures within a stretch of river is also key for many species of fish. The water temperature is critical for both breeding Salmon and Trout. The hatching success of their eggs starts to decline in water that is over 12°C and the water temperature is also key to young Salmon choosing to leave the river system and migrate to the sea. In warming rivers fewer Salmon do this, and if they don't do it, they don't mature into breeding animals that will return to the rivers to spawn future generations. The importance of tree shade to fish may not seem readily apparent to us, but as always with wildlife there exists a long-evolved balance that we can readily upset. Riverside trees do so much for wildlife, including helping to regulate the water's temperature. In a warming climate, fish like Salmon and Trout need more trees alongside rivers, not fewer.

But the most serious cause of fish stock decline is what we are putting into our rivers. Pollution from concentrated industrial farming can have very serious impacts, such as those apparent on the Wye. But as the people living alongside the Wye can testify, sewage too is a big factor. But this is not just happening in one river system. Every single day in the UK untreated sewage is discharged into our rivers right across the country. This has a whole host of detrimental effects, of course, including for those of us that use the rivers for swimming, canoeing and so on, but the principal effect is how it impacts on the ecology of our rivers. In 2024, 54% of UK waterbodies, including the Taw, failed their basic ecological targets due to sewage pollution. It would appear that these discharges don't directly impact the Otters themselves[9] – they don't seem to be suffering as a result of them – but they can affect them indirectly by reducing the amount of oxygen present in the water, which in turn drastically reduces the amount of aquatic life able to survive in those waters, be that the fish themselves or the invertebrate life they require for food. If the fish in a stretch of river die out or are severely depleted due to the impacts of sewage pollution, that stretch of river also loses its ability to support Otters. Our wild fish numbers are not declining because of Otters; they are declining for the same old reason. Us and the way we treat our watercourses.

River pollution is something that needs to be tackled and the solution on paper seems relatively straightforward: we simply need to stop polluting rivers. But what about those stretches of river that were 'improved'? Can they be unimproved?

CHAPTER 9

Bending the Line

The tree-lined Taw slips quietly underneath the railway line that bears its fictional character's name, before flowing away from human-made linear features. The sunlight filters through the canopies of the tall trees, causing the water surface to be adorned with ever-changing patches of sun and shade. Small clouds of flies dance in the sunlight, whilst a Robin eyes them voraciously from the shady recesses. The flow of the river is languid here: no rushing water, no turbulence, just a smooth onward flow. The tree-lined sides tighten briefly, the shadows seem to win, but it is a momentary victory, for the river suddenly broadens and begins to bend. Sunlight pours down onto the surface of the river's perfectly formed 'C' shape, carved into the Mid Devon countryside by the natural flow of the river. It seems to be heading back towards itself, almost forming a full circle, before it curves away again, creating another 'C' shape, but this one not so perfectly carved, as the river heads off in another direction once more.

But the direction it takes is short-lived, with the river making a sudden acute turn to the west. A large wide expanse of water forms at this bend, a broad shallow pool that hardly seems to flow, but flow it does, from the open sunlight back into the dappled shade of a narrower tree-lined section as it heads back towards the linear feature of the Tarka Line railway before slipping under it once more. A passenger on the train will have travelled 460 metres from the first bridge to the second, but an Otter, taking the riverine route, will have travelled just over 1,000 metres to complete the same journey. A natural slow lane.

After centuries of river manipulations that peaked (but did not stop) with the countrywide river improvement works in the 1970s and early 1980s, following the 1973 creation of the new water authorities, a realisation has finally dawned that these works perhaps cause more problems than they actually solve. These poorly named 'improvement works' invariably comprised of the straightening and deepening of the watercourses, effectively creating large, fast-flowing drains to move the water on quickly through the river system. The primary reason for the vast majority of these works was to supposedly help prevent flooding. But as I have already mentioned, the reality is that it only does so for that one individual section of river, simply moving the risk of flooding further downstream and greatly increasing the risk of it there in the process.

We now realise that slowing the flow of water, especially when we are increasingly experiencing more frequent intense rainfall episodes, is the key to helping reduce the risk of flooding along the whole river system. It can also be very beneficial to the river's wildlife. We can undertake some work to reverse the damage we have already done, and these works can be done anywhere along the river system. From restoring peat bogs in the uplands, where many rivers have their source, to restoring salt marsh in estuaries where the rivers finish their journeys, we are able to improve on our previous 'improvements'; we can unimprove them. These new *real* river improvement works tend to be carried out under the umbrella title of 'restoration', but personally I prefer the word 'reparation': making an amend for damage done.

Probably the most impactful of these restoration or reparation works is the realignment of rivers back to their original route – an unstraightening, a bending of the line, a recurving, a remeandering. To do this, though, takes considerable effort and resources, but it can be done and should be done where it is feasible. It costs money, but the government estimates that each year in the UK, flood damage is costing us £1.4 billion. Surely, therefore, any expenditure on repairing river systems has the potential to be very quickly repaid.

In 2012 the Royal Society for the Protection of Birds (RSPB) took over the management of 3,000 acres of land in the Lake District in the north-west of England. Within this land holding ran the Swindale Beck, a small river that had been meticulously straightened and deepened around two centuries before. The valley the beck flows through had, before it had been 'improved', often flooded during high rainfall; it was basically a natural floodplain for the watercourse, allowing the excess water to spread out, dissipating its force and flow before it slowly returned to the river's natural course as the levels dropped.

But even though the land in the valley was not great for farming, it was still land that was coveted and therefore the beck was, in the eyes of the landowners and local community at the time, ripe for being improved. The work was carried out in the early 1800s and regardless of what you think about it now, you have to acknowledge that what they achieved was pretty remarkable. There were no mechanical diggers, no towering swing shovels, no generator-powered pumps. There were just people with spades and picks. In all, these labourers' laborious work managed to straighten, deepen and widen a one-mile stretch of the beck. Huge boulders were also installed along the newly created banks of the watercourse, to help strengthen these new banks and to prevent the beck from eroding its own path once again. From an engineering point of view, it was a great feat. But from a wildlife point of view, it was terrible.

The once meandering beck and its flood meadows were transformed over the one-mile stretch into a fast-flowing torrent. The gravelly beds that would have been used by

Salmon were gone and had no chance of coming back due to the increased scouring flow of water that now roared straight and fast through the valley. The riverside vegetation had been cleared, taking with it habitat used by an abundance of species as well as the shading and therefore cooling affect it had on the water. Instead of variations in flow, variations that would produce eddies and backwaters perfect for a whole host of invertebrates such as dragonflies to breed in, as well as for fish to rest up, there was now a homogenised torrent. Otters would still have been able to use the beck, but its value to them as a territory would have been extremely limited. It is possible that they could have used the bankside boulders as holts, but the new torrent-like flow of that stretch would have made it a dangerous place for them to introduce their young to the water, assuming that there were still enough feeding opportunities present for a female to rear young in the first place.

And of course, now that the water was no longer slowed by the river's natural course, it could power on unimpeded, increasing the risk of flooding for those living further downstream – including those living in the town of Carlisle, a place that has had serious flooding problems for many years.

The RSPB – along with the modern-day landowners United Utilities, as well as Natural England and the Environment Agency – decided to, in their own words, rewiggle the river. Reintroducing the natural meanderings of a river is quite an undertaking, even when using modern-day machinery, but the work on the Swindale Beck was completed successfully and the watercourse, as it flows through the valley, now has a natural curvaceous look to it. Not only do these new curves slow the water flow, but they also increase the capacity of that section of the beck to hold water. In total, over the old one-mile length of land that the beck flowed through, the reparation works have added an additional 180 metres of length to the beck. More length of watercourse equals more capacity to hold water.

But it is not just about slowing the flow and adding length; it is also about improving (in the real sense of the word!) the watercourse for wildlife. The variance in flow has returned to the beck and so the gravel banks and bars have come back too; remarkably, within just a few months of the work's completion, Salmon were back spawning on this section of the beck, the first time they had been able to do so for two centuries. The variance in flow also provides habitat for a wide range of invertebrates that can now breed and live on the stretch, invertebrates that provide vital food for birds such as the Dipper and Grey Wagtail, and of course for a wide variety of fish. And as the beck became better for fish, so it became better for their predator, the Otter. Otters have now been recorded hunting along the new route of the river, no doubt taking advantage of the easier and increased fishing it provides. Large numbers of trees have been planted along the new banks; these will not only shade the water to help provide cooler sections that help with fish breeding, but their roots will also help stabilise the riverbank and, over time, they will potentially form natural holts in which the returning Otters can breed.

The meadows alongside the beck have once again become flood meadows. Wildflower-rich and diverse, they not only improve the overall habitat (as well as the aesthetics) but can also function as overspill sponges once more, helping to slow the flow of water and to soak up excess water in periods of high rainfall. All this in turn helps to reduce the risk of flooding further downstream.

The project has been very successful, and I am pleased to say it is just one of several such schemes of recurving rivers that have been carried out across the country. These are all schemes that should become examples to follow, as part of the much-needed national plan for our rivers that I first mentioned in a previous chapter – a plan that would help develop this sort of real improvement works into a template for the future.

We can use our skills and our technology to help us undo the damage we have done to river systems in the past; we can help Otters coexist with us. But there is no getting away from the fact that even though Otters have made a comeback, they still face challenges. And sometimes, those challenges require us to help them in a far more hands-on sort of way.

CHAPTER 10

Rescue, Rehabilitation and Release

I t is a fantastic spring day, with a blue sky, bright sunshine and real warmth in the air. I swing the car over the mill stream and roll to a stop. I get out, enjoying the sun on my back, and take in my surroundings. The trees between me and the river are all bursting with buds, their skeletal dendrite forms about to become lost in a cloak of fresh green. The air is alive with sounds; the flow of the river and the mill stream merge into a background hum, and a queen bumblebee drones past my head, her buzz lasting as she heads to some bright yellow dandelions growing on the margin of the hardstanding. A Chiffchaff is singing its discordant but still beautiful song, a Robin gives an accompanying short burst and a Wren blasts out its own contribution. Several Blackbirds are singing from all directions and a Song Thrush repeats itself in the distance.

Above me, in the uninterrupted blue, Rooks whirl in the freedom that only birds can know. I look down the broad valley laid out in front of me: the oaks that clothe the rising sides are also coming into green, their new leaves beginning to coat the woodland, and further off the darker forever green of towering Douglas Firs stands out, dominating all other trees as they reach skyward, true giants, yet still with many metres of procerity to potentially add.[1] But I am not here to admire the sylvan life of this part of the Taw. I am here to see Otters, and lowering my gaze to what immediately lies in front of me, I see where I am going to find them. And I use the word 'them' deliberately, for there is not just one Otter here, and not two or three, there are several. You could even say there are too many.

I am at the beautifully located base of the United Kingdom Wild Otter Trust (UKWOT). It is the largest Eurasian Otter rehabilitation centre in the world and lies in the valley of the River Taw, right in the heart of Henry Williamson's Two Rivers Country, but it is a site that has more connections to *Tarka* than just being located in the Taw valley. The buildings I stand beside were used as a location in the film *Tarka the Otter*, directed by David Cobham. The film, which was released in 1979, was filmed in 1976 and 1977, and this site was used as a location. I am stood in Tarka's film set.

I am here to meet Dave Webb, the founder, CEO and all-round main man of the UKWOT. When it comes to Otters, there are no more dedicated people than Dave and his wife Cathy. This site is not exactly secret, but nor is its location publicised. This is not somewhere for people to come and visit, but a site to help Otters in need. It is a beautiful place and the work carried out here is brilliant, but I find myself wishing that it did not need to exist. But the realities of the world in which we live mean that it does.

There are currently seventeen Otters in residence, all of which are at various stages of their rehabilitation. In the month leading up to my visit a further ten Otters have already been released back into the wild following their successful stay here. This is what the site is about: rescuing Otters, caring for, nursing and tending them, rehabilitating them, and then releasing them back into the wild where they belong. This is the reason the public are not asked to come and visit – the animals here cannot get used to humans. They are wild animals when they arrive and Dave and his team ensure that that is how they stay, before they are released back to their natural habitat once more.

Dave's journey to this point started when he was just 15; he should have been in school, but as he freely admitted to me, he was skiving off, enjoying the outdoors rather than the confines of the classroom. Instead of being in a lesson he found himself having a chance encounter with an Otter, and that encounter fired a passion within him. But as the story often goes, life happened and his passion, whilst still aflame, got put on the back burner. That is until he moved to Devon in 1998 and had yet another ottery encounter, this time with a female and her cub. The passion was well and truly reignited.

Dave founded UKWOT that year and initially worked with anglers and fisheries, helping them to manage and resolve the conflicts that we looked at a couple of chapters ago. He soon realised that there was plenty of work, and in 2016 Dave registered UKWOT as a charity before opening his first rehabilitation centre the following year. It quickly became obvious that there was, sadly, a demand for such a place, a demand that soon made Dave realise that he needed a bigger site. Thanks to the generosity of landowners and donors Dave and the UKWOT moved into their current location in 2021. So far, the charity has rehabilitated and released back into the wild well over 200 Otters, a phenomenal achievement. Dave is very proud to report that every rehabilitation and release has been completely successful. The Otters that have stayed in this centre and subsequently been rehabilitated and released have only ever been here once; there has never been a need for a readmission.

We walk down to the pens and enclosures. The first two of these hold the only permanent residents, two Otters that are simply too tame and habituated to humans to leave. One of these was rescued as a cub and taken to a wildlife hospital, but during its care there it became too dependent on the humans who were looking after it to ever be released back into the wild. UKWOT stepped in when the wildlife hospital was faced with the dilemma of having to have the now healthy animal euthanised, as they no longer had the capacity to care for what was now essentially a pet Otter. The other permanent Otter resident here was most probably kept as a pet, but the story behind its origins is unknown. Dave recalls when he first met it; normally, he tells me, when he goes to rescue an Otter the animal in question either tries to get away and hide from him, or it attacks and tries to bite him, but this Otter didn't do that. At their initial meeting, Dave sat down near the animal to see how it would react to his presence. The Otter casually wandered up to him, put its front legs on his legs and pulled itself up onto his lap where it then promptly settled down, much like the proverbial lap dog. It was very clear that this was an animal that was anything but wild.

However, the two resident Otters, the ones that are used to people and therefore the ones I am most likely to encounter on my visit, are apparently not interested in seeing me. The warmth of the day no doubt contributes to their decision to both stay slumbering in the cool darkness of their boxes. I ask Dave where the wild Otters in his care come from, and what the main reason is for them ending up here. He tells me that by far the primary cause is them becoming separated from their mothers. Cubs can be dependent on their mothers for up to a year – they need that maternal care to survive, and if they lose it, especially in the early months, then they are in trouble. Most of them will die, but a few are lucky; they are the ones that are found and rescued.

The main cause behind cub and mother separation is, Dave believes, road casualties: the mother gets hit on the road and the young cubs are left alone. It is an all-too-common occurrence; we have already seen in a previous chapter that significant numbers of Otters are killed on our roads every year. Whilst we do not know what percentage of them are females with cubs, it is likely a high number. Although these road traffic accidents are a common cause, road casualties are not the only reason Dave has come across for young Otters becoming separated from their mothers. The mother Otter can perish in all sorts of other human-induced ways. Sadly, illegal persecution is one of these, as is the adult animal getting herself trapped in underwater nets and drowning. Whatever the cause of the sudden separation, the future for the young cubs left behind is often short. If the cubs are old enough to navigate the territory in which they were being reared, they will have more of a chance of being discovered and rescued – the young animals hopefully being noticed by the public, as they call ever more frantically for their parent to return, their weakened state leading them out into the open where they are much more likely to be seen. Whilst some are indeed spotted, the vast majority of separated Otter cubs will not be noticed. Only a lucky few are, and that's where Dave and his team at UKWOT come in.

As soon as Dave gets the call that a young Otter is obviously in distress, he drops everything and heads off to check on it, wherever it may be. If the Otter needs help, then the next thing he has to do is catch it. This is not necessarily the easiest of tasks, but when in a weakened state and acting all helpless alongside a watercourse, the young Otters can be relatively quickly captured and attended to. Once secure, they are taken back to the centre and the long road of rehabilitation begins.

But not all the Otters that are being rehabilitated at the centre at the time of my visit have been found helpless alongside watercourses searching for their mother. The next pen we come to is home to two Otters, and these two are not in their box but out and about. I am seeing Otters by the Taw and I am very happy about it. But these Otters are not so happy to see me; they run, with the typical arch-backed gait of these mustelids, into the cover of their enclosure, disappearing for a moment before the lure of what they think might be a possible feeding opportunity piques their curiosity. Out they come, one after another, into the bright sunshine, showing me their full beauty before they quickly lose confidence and vanish underneath the raised concrete base on which their box is placed. Two muzzles poke out from underneath it, their small, somewhat beady eyes just visible in the shadow as they observe me nervously. I am thrilled to have seen these beautiful animals at such close quarters, and Dave is thrilled that they have behaved as they did: hiding away, keeping out of sight, whilst trying to ascertain what is happening. This is behaviour that they will need when they are eventually released.

But my delight at seeing them quickly turns to anger and sadness when I hear the story of how they ended up here.

These two beautiful creatures were found deliberately dumped in a crate next to a road in Shropshire. They were severely underweight, soaking wet and very lethargic. The cubs were close to death when, by sheer chance, they were discovered by a member of the public; that person's prompt action saved the lives of these two beautiful creatures. How these two very young cubs ended up in a crate alongside a road is likely never to be known, and how it was that they became separated from their mother is, again, probably to remain a mystery. What can be ascertained is that the person responsible for dumping these two animals was evidently very cruel and very cowardly. As Dave said at the time of rescuing them, 'this is a new low … a planned act of unprecedented cruelty'.

Six months later and I am pleased to say that they are now doing well, thanks to the work of Dave and his team of dedicated volunteers. Their rehabilitation is on track and in another five months or so they will be released back into the wild once more. We like to think that in the hundred years since *Tarka* was written, we have moved on when it comes to the cruelty we are prepared to inflict on animals – but as this incident showed, not everyone has done so.

To rehabilitate an Otter, to pay for veterinary treatment, to house them and feed them for up to a year costs money. Dave estimates that on average it costs £3,500 per Otter to do so. I look around me at the various enclosures. There are seventeen

Otters being rehabilitated at the moment; another ten finished their rehabilitation and have been released in the last month. It takes a lot of money to do what UKWOT are doing, and all of it comes from the kindness of people who make donations to their cause.[2]

I ask Dave about the future, about what issues he sees as becoming more problematic for Otters, the issues that will lead to more of them that need his help. He starts off by saying that he thinks, despite the case of the two Otters above, that deliberate persecution has dropped and whilst there will always be some persecution, it is now much lower than it once was. He knows that fisheries in the past have been tempted to take the law into their own hands when it has come to dealing with an Otter that is causing them a problem; but now with improved and widely known fencing techniques, this threat has greatly reduced. He also thinks it has reduced because he has a government licence to humanely live-trap Otters that are causing problems for fisheries. Businesses that may not have known what to do otherwise can contact him and his team of advisors for help. There is now an alternative to making the wrong and illegal choice.

The risk that Dave sees as increasing in the future, though, is one of our own making, or rather building. Wherever you go in the countryside there are new developments being built, industrial complexes, new road schemes and new housing estates, often with road names that give an ironic nod to what was there before the development. I know one new housing development that has even been named Tarka View. As I write this, the government's proposed Planning and Infrastructure Bill is making its way through Parliament. It is a bill that undermines species and habitat protection across England, pushing the presumption that planning permission will be given approval despite concerns about how it will impact wildlife.

It is this that Dave sees as the biggest risk to Otters in the future. Increased development leads to increased fragmentation of habitat – a new housing estate or business park is a very effective barrier for any young Otter needing to disperse from its own river system, and of course any new roads, be they residential or bypasses, have the potential to cause yet more Otter road casualties. As we will see in the next chapter, the recovery of Otters in the south-east of England has been slow and this development and infrastructure growth may well be playing a key part in that sluggish rate of recovery. Dave finds this new proposed bill very frustrating, and can foresee the potential damage it may bring.

He is not alone in this. Numerous wildlife NGOs have written to the government expressing their concerns; 32 CEOs of various organisations have signed the letter pointing out that unless the act is significantly amended it will break new ground in the destruction of nature across England.[3] Dave hopes that common sense will prevail, but he is not optimistic. UKWOT have recently acquired another four acres of land to expand their rehabilitation work, such is the demand already for the work that they do. How much more land will they need in the future?

As we talk about the future, we also talk about climate change and how that may well increase admissions to the rehabilitation centre. As we have already seen, Otters are more at risk of getting hit on our roads in the winter months when there is increased rainfall and the rivers are running higher, making passing through structures such as culverts and bridges a hazard that many Otters seek to avoid.

Increased torrential rain in the spring and summer months inevitably leads to our rivers running higher and faster; many human communities have seen this for themselves with flash flooding causing misery and financial loss. I ask Dave if he thinks these unseasonal downpours will lead to an increase in admissions to his centre. Not only does he think they will, but he also believes he is already seeing evidence of it happening.

When Dave first started out rescuing, rehabilitating and subsequently releasing Otters, the work was very seasonal.

The typical time for admissions to the centre was during the winter period, from October through to the end of February, a period that fits well with the data from the Road Lab study we discussed earlier. Now, however, Dave has admissions in every month of the year and they are heavily linked to spells of increased rain. We talk about the current year as an example: January was very wet, with all of England, on average, receiving 116% of the long-term average amount of rainfall for that month. In other words, it rained more than normal. This increased rainfall led to an increase of admissions for Dave and his team, with the centre taking in nine animals in just that month alone.

But from mid-February onwards, the rain stopped falling in many parts of the country. The month of March was very dry indeed, with England, on average, only receiving 22% of the long-term average amount of rainfall for that month. Unsurprisingly, the number of new admissions dried up along with the weather. It is mid-April when we chat, and over the last eight weeks there have only been three admissions. Dave knows that if the weather breaks with a spate of prolonged heavy rain, those admissions will once again rise in tandem with the river levels. Whereas in the past Dave would gear up and prepare for the 'admissions season' of winter, he and his team now have to be ready throughout the year as our climate, and in particular our rainfall, becomes ever more unpredictable.

We do not see all the Otter enclosures; it is important that human contact with the rehabilitating animals is kept to a minimum. The Otters move through the site as their rehabilitation progresses, shifting ever further down the line of enclosures, each one that bit further away from the buildings. As they move further away from these, so their contact with humans reduces – as they recover from their ordeal, their need for human intervention such as veterinary checks also reduces. By the time they reach the last enclosures they are being prepared for release back into the wild, and the last thing they need is unrequired human contact. I look towards the last enclosures, knowing that they contain animals that will soon be back where they belong, in our rivers. I leave Dave to prepare for their imminent release and follow the road back along the Taw valley.

On my homeward journey I think about the brilliant work I have just witnessed and how people like Dave can make a real difference. A bit like the unbending of rivers, we can do very positive things for our Otters. But as I drive alongside the Taw, I am also acutely aware of something potentially sinister lurking in its waters, something that could well have very serious implications for the Otters, for other wildlife and for us. I am thinking about what we are putting into the water every single day.

CHAPTER 11

Lessons Learnt or Forever Failing?

White-topped waves form moving ridges of water out in the bay, the stiff breeze forever playing with their integrity, causing continuous cycles of cresting and collapsing. I am looking out at Bideford Bay, a part of the Bristol Channel, itself a large inlet of the Atlantic Ocean. In the distance I can just make out the hazy lump of Lundy, Devon's island outpost, the choppy waters around it making it look as if it's the island that moves up and down, rather than the sea. It looks to my non-maritime eyes like a tumultuous environment to be in – the swell, the wind and the various currents causing continuous random disturbances on the water's surface. I am watching all this play out in front of me from the safe vantage point of the north tip of Northam Burrows, just to the west and north of the confluence of the two rivers of Two Rivers Country. I am glad that I'm stood on the dry sandy ground, but not everyone is. A small fishing boat is making its way back in towards land, heading for the mouth of the two rivers, its dark blue hull bobbing along in a manner that seems to me alarming. But the bright oilskin-clad human figure stood on the side of the boat seems to be completely unbothered by the movement of the vessel. Another load of locally caught fish is heading for the quay.

The scale of devastation to our wildlife in the latter half of the last century, caused by the use of organochlorine and organophosphate compounds in agriculture, was quite simply shocking. It was such a trauma that the authorities had to respond; the chemicals were withdrawn and then eventually banned. The whole of the country has been affected by the use of these synthetic chemicals, humans included,[1] and their liberal usage across large swathes of land has led to our rivers becoming conduits for them. As we have seen, Otters, at the top of the riverine food chain, were particularly badly affected, the result being that they were extirpated from most of England, almost completely dying out as an English mammal. Mistakes will always be made, but mistakes can and should be learnt from, so have we learnt from this catastrophic mistake? Are our rivers now clean and healthy? Have the poisoned arteries been purified?

One word can answer all three questions. No.

Watershed Investigations is a public interest journalism organisation that aims to shine a light on all aspects of the water crisis. In particular, the organisation is very hot on water pollution and its causes. Watershed have produced a publicly available interactive map of Britain and its watercourses, called, logically enough, 'The Watershed Pollution Map'.[2] There are many layers to this map, each of which deals with a specific issue; I decided to use the layer that shows ecological and groundwater health testing results for the Taw. Bearing in mind that the river starts high up on Dartmoor and then flows through the bucolic countryside of Devon, I thought it would be a fairly clean river. I was wrong.

For the purposes of the map, the river is split into five sections covering its entire length from source to sea. Each of those sections is graded on its ecological classification and its chemical classification. For each section of the river, the ecological classification of its waters was rated as moderate – the Taw is not in good ecological health. But it was worse when looking at the chemical classification: every one of the five sections was rated as failing when it came to their chemical content. Shocked at this result, I wondered if it was perhaps an anomaly; perhaps for some reason the Taw is a particularly bad river in this respect. To test that theory I started to look at other Devon rivers, and the results were pretty much the same. I then spread my search wider, randomly zooming in on other areas of the country. It soon became obvious that the Taw is not a particularly bad river – it is, sadly, a very normal river. Our rivers may no longer be full of the agricultural chemicals that did so much damage in the second half of the previous century, but they are in no way clean. Some chemical pollutants may have gone, but we seem to have a terrible knack of replacing them with new ones.

Perfluoroalkyl and polyfluoroalkyl substances are, thankfully, normally just referred to as PFAs. They are a very large group of synthetic organoflourine compounds and just like the organophosphates and organochlorines of the last century, they are not great things to introduce into the environment. Unfortunately though, these compounds are already here, and once they arrive, they are very good at staying. PFAs are often referred to by another name: forever chemicals. It is a very apt description, because these human-created chemical compounds will be with us for thousands of years. They are not all new creations; they first appeared in 1938 when they were used in the manufacture of Teflon for making non-stick cooking utensils. Since then more and more of these compounds have been created and they have been used in more and more types of products – everything from the nylon we wear, to the nail polish and make-up we apply to ourselves; from food packaging to fire-fighting foam, the use of these forever chemicals has become routine. As their usage diversified and intensified, we began to realise that these products might not be that good for us. There has been some small-scale bolting of the now empty stable's doors, resulting in some of the PFAs being banned, but there are estimated to be around 10,000 different compounds and most of them are still in use today. We now know that

PFAs are heavily linked to a variety of cancers in humans, as well as issues in human foetal development and birth defects. It has been suggested that due to their widespread use and persistence, traces of PFAs can be found in the blood of every human on the planet.[3]

When they were first produced, PFAs were thought to be chemically inert. Now we know they are not. We also know that they move through soils with ease and are found in rain, drinking water and waste water, all three of which end up in our rivers. They bioaccumulate in wildlife, especially in fish, moving up the food chain as these are eaten by predators and humans alike. It all sounds depressingly familiar, doesn't it? Lessons most certainly not learnt.

Returning to the Watershed Pollution Map, I click on the PFAs layer. Straightaway I notice two coloured circles in Bideford Bay, close to the mouth of the Taw/Torridge estuary, the very place the small fishing boat at the beginning of this chapter was navigating. The first of these circles relates to the carcass of a Harbour Porpoise that was found washed up on the shore. The second relates to a random sample of fish species caught by fishers and analysed later. All the fish in that sample, as well as the Harbour Porpoise (a predator of fish), contained levels of PFAs in their tissues.

PFAs enter our river systems from a variety of sources. They can leach from landfill sites, for example, but the greatest source is via waste water, both domestic and industrial. This is processed at waste water treatment plants, often referred to as sewage works, but these are not equipped to deal with PFAs; they cannot filter them out or break them down. PFAs that enter a waste water treatment plant are able to pass straight through them in the water that is then discharged into our waterways as treated effluent. There is nothing stopping their passage into our river systems.

But that is not the only way they pass through our waste water treatment works – they also do so in more concentrated and solid form as well. The solid waste that is filtered out of the water at sewage works is often referred to as 'sewage sludge' (although there are some that prefer to give it the nicer-sounding name of biosolids – can't think why they don't want to call it what it is…). We produce a lot of sewage sludge, as you might expect; in fact the UK produces around 3.6 million tonnes of it every year. It has been reported that 87% of this, or 3.13 million tonnes if you prefer,[4] is then routinely spread on our agricultural fields as a fertiliser, a fertiliser replete with concentrated PFAs that subsequently wash down the fields in the next bout of heavy rain and head straight into our river systems. Water UK, the trade body of UK water companies and waste water processors, says that these biosolids are spread onto our agricultural fields because they are a soil improver that offers an important source of nutrients and organic matter for growing crops. Conversely, the James Hutton Institute suggests that the use of sewage sludge on agricultural fields could be contributing to poor soil health.

Whichever of those two standpoints is correct, the spreading of sewage sludge on our agricultural fields happens all the time; it is a routine part of farming. I see it regularly around the watershed of the Taw, often on the day before heavy rain is forecast. It is a countrywide practice in areas of intensive farming; it is the norm. It is also the main route for the forever chemicals (and as we will see, so many other chemicals) to enter our watercourses, which we humans use for bathing, fishing and so on. And of course, watercourses used by Otters.

What with PFAs having multiple and easy routes into our rivers and the fact that the compounds bioaccumulate in fish, it should therefore come as no surprise to learn that PFAs have also been found in Otter carcasses. A study carried out by the Otter Project based at Cardiff University examined the livers of 20 dead road casualty Otters. Every single one of them had PFAs present in their tissues. The European Union, but not the UK, are considering setting an environmental standard limit on the amount of PFAs that fish can have in their tissues, to prevent people and predators such as Otters from bioaccumulating the compounds in their own bodies. The limit the EU is considering is 0.077 µg/kg (0.077 micrograms per kilogram). In one recent study in the UK carried out by Watershed Investigations in collaboration with the Marine Conservation Society and the *Guardian* newspaper, a liver from an Otter roadkill carcass was found to contain 9,962 µg/kg of PFAs.[5] That is nearly 130,000 times the limit the European Union is saying should be the environmental standard for fish. It would appear from that data that any future imposed standard is simply going to come far too late.

In 2012 a PFA known by the acronym PFOA was banned in the EU and therefore in the UK as well. It is the PFA used in the process of making Teflon; previous studies by Cardiff University's Otter Project had shown that there were high concentrations of PFOA in dead road casualty Otters recovered from areas near a factory in Britain that used them in the manufacture of its products. Since 2012 that factory has stopped using the now banned PFOA, but despite this ban, the persistence of the chemical compound in the environment has been clearly demonstrated, with it still being found in the bodies of roadkill Otters over ten years later. However, it is not the only PFA being found in the carcasses. New PFAs, created to be 'safer' replacements for the banned and/or regulated ones, are also now being detected in the tissues of the dead Otters. But what is more alarming is that other PFAs that are not even used in Britain are also being discovered in these dead Otters.

PFAs have a global reach – 19 out of 20 Otter carcasses sampled by the research team had detectable levels of a PFA called F-35B. This chemical compound is used in China in the electroplating industry; it is not used anywhere near Britain, but it is being found in our Otters (and would no doubt be discovered in other wildlife if it was looked for). This can only come about due to the importation of products made in China that contain this chemical compound, and their then subsequent disposal in our landfill sites. PFAs leach into the environment with ease, and once they do, they stay there, making their way through soils and into our river systems. We may ban certain PFAs, we may not even use them, but the modern global economy ensures that they are still found here.

The knowledge that PFAs seem to be routinely discovered in the tissues of roadkill Otters is troubling indeed. PFAs are not only known to be linked to cancers in humans but also to pregnancy issues and problems with foetal development. What made the previous poisoning of Otters (and many other species) in the 1950s and 1960s so insidious was that before it reached levels that would kill an animal outright, it badly affected their breeding capability, causing the populations to fall precipitously without alarm bells being raised by the presence of actual dead animals. Are we about to witness history repeating itself? Henry Williamson's view of Otters was probably different to my own – we certainly would not have agreed on hunting them – but I think he would be as equally depressed about the prospect of yet another chemically induced population crash for these riverine mustelids.

And as we approach the one hundredth anniversary of *Tarka the Otter* being published, there is evidence that something is perhaps going wrong once more with our Otters.

The National Otter Survey of Wales is carried out periodically to monitor the Welsh Otter population. The fifth Welsh survey was carried out between 2009 and 2010, and, just like the corresponding survey in England, it demonstrated the continuing comeback of the Otter since the pesticide-induced nadir in the latter half of the last century. In all, over a thousand sites were surveyed and a pleasing 90% of them provided positive records for the presence of Otter.

This survey was repeated between 2015 and 2017 in the sixth Welsh survey; the methodology was the same as in the fifth survey, but the results were not. They showed a substantive decline in the number of sites providing positive records of the presence of Otters. In less than a decade, the pleasing figure of 90% of sites with Otters present had fallen to a less encouraging 70%. That 20% drop in positive sites is an average across the Welsh river systems as a whole, but some of these river systems showed as much as a 26% drop in the number of sites producing positive signs. So potentially, over a quarter of all the sites in those river systems surveyed in the fifth survey have subsequently lost the Otters that were present there less than ten years ago. That is an alarming drop indeed.

But what of England? Like the fifth Welsh survey, the English fifth survey had shown a continued increase in the number of sites with Otters present. Would the sixth survey show this increase continuing, or would it too show a worrying decline? The English sixth National Otter Survey was carried out during 2022 and 2023 by the Mammal Society, repeating the methodology used in the previous ones. As this book is being written, the results of that survey are still yet to be published, but I was fortunate enough to see and discuss them ahead of their publication with the Mammal Society's CEO Matt Larsen-Daw.

My talk with Matt was a very positive one, both in how we got on and, I am relieved to say, what we discussed. The upward trend in Otter numbers in England has continued, in all the regions and all the counties. In the fifth England survey, 56% of sites surveyed produced positive signs of Otters. The results of the sixth survey show that number has climbed once more, and it is a significant climb, to 75% – very positive news. However, the recovery of the Otter in the south-east of England is still very, very slow with only between 11% and 25% of sites across the whole region showing positive signs of Otter; and in Kent, the last county in England to be recolonised, the recovery is particularly glacial. The reasons are not completely clear, but it is highly likely that fragmentation of habitat due to development and new infrastructure, as well as the naturally fragmented network of the river systems themselves, coupled with a high density of busy roads, is an important limiting factor hindering the full return of the Otter to this part of England.

For me, a big positive to come out of my chat with Matt was the fact that the region known as the Welsh Borders had one of the highest positive results, with over 90% of sites surveyed showing positive signs of Otter presence. As the rivers in this area, such as the Wye, come from across the border in Wales, it perhaps indicates that the previously mentioned drop in numbers in the last Welsh survey are not as bad as they initially looked. The seventh Welsh survey is in progress as I write this, so hopefully the results from that will show that the numbers in the sixth were an anomaly and they have at least stabilised or are hopefully climbing once again.

Whilst the results of the sixth English survey undeniably demonstrate that the positive trend in the Otter population of the country is continuing, there are still serious concerns that need to be considered. There are a number of clouds on the horizon, and they are being caused by pollution.

Surely the biggest pollution cloud on the horizon for Otters are those forever chemicals, the PFAs. In my chat with Matt Larsen-Daw, he mentioned them as a growing concern for all involved in Otter conservation. The results of the sixth national survey indicate that, at the moment, the accumulation of these forever chemicals in the bodies of Otters (the Otter Project study previously mentioned found them present in 100% of the Otters tested) is not apparently impacting on their populations. But that doesn't mean that they won't do so in the future, perhaps the near future. Whilst many countries, including those in the EU, have stringent regulations in place over the use of PFAs and how they subsequently contaminate the environment, Britain currently does not. Every day sees more of these compounds making their way into the river systems of this country; are these river systems becoming poisoned arteries once more?

There is already some evidence that PFAs are having a detrimental effect on the breeding of the North American River Otter, a close relation of our own Otter, but it should also be pointed out that this species exhibits something known as delayed implantation. This is a breeding strategy used by several members of the mustelid family including our own Stoat and Badger, but not our Otter, whose breeding biology is therefore different from that of the North American River Otter. Our Otter is mated and the female's eggs are fertilised with the resulting embryo implanting into the female's uterus, where it develops constantly from that point until birth a couple of months later. But in species that exhibit delayed implantation, the eggs are fertilised during the act of mating but then do not immediately implant into the female's uterus. Instead, their development is paused until a physiological trigger causes them to become implanted into the uterus at a later date, from which point they continue their development.

One of the advantages of this stratagem is that, though where a species is widely spread and living at low densities, the animals can breed at any time of the year, the births are timed to happen at an advantageous period such as during better climatic conditions or when a particular food source is abundant. The length of delay, known as the diapause, in the implantation and subsequent full development of the embryo varies between species. One mustelid, the Western Spotted Skunk, has a diapause of 200 days; but another, the American Mink, has a diapause of about 14 days. As it is a physiological trigger that determines when the embryo implants, the length of the diapause can vary a lot within a species. The North American River Otter has a diapause that can last between seven and nine months – roughly speaking, between 200 days and 270 days. The large range in the number of days is most likely related to the latitude at which the animal lives, with those in more northerly areas experiencing harsher winters than those that live in more southern climes. Those in the north have a longer diapause, reflecting the longer time it takes for the harsh winter to end and for more favourable conditions in which to rear young kits to arrive.

PFAs have already been shown to negatively impact the fertility of American Mink. As already mentioned, this is a species that exhibits delayed implantation, so perhaps this breeding strategy is more susceptible to harm from PFAs. But then again, perhaps not. We simply have no way of knowing what effects these forever chemicals are going to have on all of our wildlife species, regardless of their breeding stratagem. PFAs are dangerous; scientific studies have already demonstrated that they are very bad for us, and we should never forget that we are a mammal species just as an Otter is.

As Matt said to me, we just do not know what the accumulation of PFAs within Otters is going to do to them. But if there is a threshold at which they start to impact the breeding of our Otters and that threshold is reached simultaneously across the country, then the Otter population could once again plummet. In answer to my earlier question a few pages back: no, we have not learnt our lesson.

In fact, we have completely ignored it. PFAs are a problem, and they are likely to become – though I sincerely hope not – a very serious issue for Otters and other wildlife. But there are other pollution clouds on that horizon; in fact there are quite a lot of them, and each one has the potential to cause real harm.

Plastic population

As part of the sixth National Otter Survey, the Mammal Society also conducted another random study. They organised the collection of a small sample of spraints gathered from a small sample of river systems during the actual population survey. This study was looking for the presence of microplastics.

Microplastics is a term given to a very modern form of pollution. To be classed as a microplastic, the piece of plastic must measure less than 5 mm in length. That's actually quite large, but the range in sizes goes all the way down to 1 nanometre, which is very, very micro indeed.[6] There are two types of microplastic, primary and secondary. The first of these relates to plastics that were already micro before they entered the environment; a great example of this type of microplastic is the glitter that often adorns greeting cards. Secondary microplastics are ones that have been degraded after entering the environment, worn down into smaller particles by a variety of processes. Sources for this sort of pollution are virtually endless but include everything from plastic drinks bottles to fishing nets, and car tyres to teabags.

For microplastics to be present in the spraints of an Otter, the animal will have had to ingest them. Shockingly, but perhaps not surprisingly, microplastics were found in 23.3% of the randomly selected spraints sampled. If that figure can be extrapolated out, and there is no reason to suggest it cannot, then a quarter of all our Otters are ingesting microplastics via their diet. But what I found more shocking was that the river system that provided the spraints with the highest levels of microplastics was not one I would typically class as an industrial river. Apart from its estuary, it is a very rural river. It is also a river that borders Williamson's Two Rivers Country, although it is not directly within it – for it is a river in Devon, the River Tamar.

The plastics found in the spraints of Otters on the Tamar were from four different sources: biochemical plastics, nylon, cellophane and polynorbornene, a plastic monomer used mainly in the rubber industry. Sadly, a quick review of scientific papers shows that not only is the Tamar not alone in its Otters having microplastics in their spraints, but the percentage of Tamar spraints containing these modern-day pollutants is actually rather low when compared to elsewhere. A recent study in Ireland produced an even higher figure;[7] as with the Mammal Society survey, a small randomly collected sample of Otter spraints were tested for the presence of microplastics. For this study the spraints were collected from eight different river catchment areas. The methodology was the same but the results were different, because the percentage of spraints containing microplastics was much higher, with 57% giving positive results for their presence.

But that number is worryingly trumped again by another study, which collected Otter spraint samples on an Alpine river in central Europe, the river Inn.[8] A total of five sample sites were chosen along the full length of the river: one in Switzerland where the river rises, two in Austria through which it flows, and two in Germany where the river meets and joins the Danube. At each of the sample sites ten spraints were collected for analysis. A total of 50 spraints were therefore tested, and every single one of them, 100%, had microplastics present, even the spraints from the sample site close to the source of the river high up in the Alps.

Otters are not ingesting these microplastics directly; they are doing so via the ingestion of their prey, like the PFAs already mentioned. The microplastics find their way into our river systems and then they accumulate up through the food chain, ending up in the bodies of the top riverine predator. And one of the ways in which microplastics enter our river systems is once again via the application of sewage sludge onto our agricultural fields.

There is legislation of sorts in place in Britain to supposedly prevent pollutants in sewage sludge from being thus applied, but the list of pollutants that are tested for is very small and nowhere near comprehensive enough. The pollutants that the tests look for are heavy metals such as lead and cadmium, neither of which we would want entering the environment, but the tests do not look for PFAs or microplastics, simply because the waste water companies are not obliged to by law. It is amazing to think that something as toxic as PFAs are not covered at all by the legislation, but even if they were, there is increasing evidence that testing even for the heavy metals is often not done at all. Based on the results of one thorough study by the University of Exeter of one British waste water plant, it has been estimated that the fields of Britain could be receiving the equivalent of 20,000 bank cards' worth of plastic every single month via the application of sewage sludge.[9] Another study has shown that in a field where sewage sludge was applied for four years, the amount of microplastics present in the soil there rose by a staggering 1,450%; and in another location the amount of microplastics present was virtually the same 22 years after they had been originally put there.[10] Microplastics are like the worst house guest ever. Once they arrive, they never seem to leave.

Other countries in Europe take this matter far more seriously than we do, limiting the usage of sewage sludge and imposing far more constraints on what that sludge can contain. The fact that we in the UK do not do this has led the Environmental Investigation Agency (EIA) to state that microplastic contamination from sewage sludge in the UK could be the highest in Europe.[11] Successive British governments have failed to manage our widespread use of this sludge, despite regulations being put in place in other European countries where the hazards have been recognised; indeed, we have looked to 'benefit' from these tighter regulations. When the Dutch government banned the use of sewage sludge for agricultural use in 1995 because of their concerns for human and environmental health, we took the opportunity to import their sewage sludge, so that we could apply it to our fields!

We do not know what the long-term effects of microplastic pollution are going to be in relation to our own health. We certainly have no idea what effects they will have on predators like the Otter, but it is not likely to be good. It is extremely troubling that the three small studies I have quoted show that not only are microplastics present in Otters, but they are potentially present in very high numbers.

Every aspect of our lives, everything we do, seems to be irrevocably linked to our watercourses. Whether we are consigning to landfill an electronic product we no longer use, driving our car, washing our clothes or our dishes or ourselves, whatever we do in our normal daily routine, some part of it will end up in our rivers. But those rivers are remote to many of us. We do not think about them day to day, we do not see that how we live our lives impacts them and their wildlife. We see and read headlines about sewage in rivers, but we do not make the connection to the original producer of that waste, us. When we consume something, some of what we are consuming will end up in our watercourses. The trouble is, we do not just consume food and drink; we also consume drugs.

Medicated mustelids

The results of a year-long survey that ended in March 2024 revealed that just under three million people in England and Wales had consumed an illegal drug in the previous 12 months. In 2023 over 1.8 billion prescription drugs were dispensed to be consumed in the UK. In that same year, we spent a staggering £3.2 billion on over-the-counter medication. Whether we obtain our drugs illicitly, via prescription or over the counter, we consume an awful lot of pharmaceuticals.[12]

Whenever we take any drug, we excrete some of it when we subsequently go to the toilet, and inevitably what we excrete will end up in our watercourses. This may, though not always, be a mere trace of the drug – but with nearly 70 million people in the UK, those traces become more concentrated. Whilst we are busy drugging ourselves, are we also busy drugging our rivers, and our Otters?

The University of York and the Rivers Trust released details in 2024 of their study looking at the presence of pharmaceuticals in a sample of rivers in the UK. These weren't rivers passing through major industrial areas, but within our supposedly most pristine of landscapes, our National Parks.[13] A total of 54 locations across England's ten National Parks were sampled, and of those, pharmaceuticals were found at 52 of the locations. To phrase it in terms of statistics, 96% of river sample sites across our ten National Parks contained pharmaceutical products that simply shouldn't be there.

The drugs that were being found included antidepressants, anticonvulsants, antimicrobials, anti-inflammatory medication, lipid regulators and type 2 diabetes treatments, to name a few. The most frequently encountered substances known as active pharmaceutical ingredients (APIs) were caffeine, carbamazepine (an anticonvulsant used in epilepsy medicine), metformin (a drug used to treat type 2 diabetes), fexofenadine and cetirizine (both of which are antihistamines). These APIs were present in over 60% of the samples taken.

The authors of the study point out that previous research looking at pharmaceutical pollution of our waterways had concentrated on sections of rivers that flow through urban areas, and that there had been a general assumption that rivers in more rural areas would have lower levels of this type of pollution. Their study shows this not to be the case; some rivers in the Peak District and on Exmoor actually had higher concentrations of pharmaceutical pollutants than river samples taken from London. According to the researchers who conducted the study, some of the levels detected were high enough to raise concern not just for the wildlife of the rivers, but also for any humans who came into contact with those rivers.

Another study carried out by Imperial College London and the environmental charity Earthwatch analysed thousands of samples taken from watercourses all across the UK.[14] This study revealed, among many other pollutants recorded, widespread levels of pharmaceutical products in our watercourses. Drugs such as antidepressants, antibiotics and painkillers were recorded in significant proportions of the samples collected.

The authors of the first study conclude that their results highlight the need for far greater monitoring of our rivers and far tighter regulation. They also say that a multi-pronged approach is required to deal with the pharmaceutical pollution problem and that the government, local authorities and the water industry must come together to provide investment in water treatment technology, with a view to reducing the amount of pollution reaching our rivers, as well as to investigate the potential impact of such pharmaceutical pollution. In other words, we should be doing something and we should be doing it now.

Currently, we simply do not know what impact all this pharmaceutical pollution is having on our river systems and the wildlife within them. Are these substances having a detrimental effect directly on Otters, or indirectly via affecting other species in the food chain? A study examining a river that flows through the Glastonbury Festival site in 2021 demonstrated that the waters, a few days after the festival, were polluted with 'dangerous' levels of MDMA (more commonly known as Ecstasy) and cocaine, both of which, the report concluded, entered the river via festival-goers' urine. The researchers stated that the levels detected could cause considerable harm to the freshwater Eels present in the river. Eels are of course a favoured prey item of Otters.

There have been very few published studies on what effects, if any, these pharmaceutical pollutants are going to have on Otters. But we do know there is great potential for many of these drugs to bioaccumulate up through the food chain, making it very likely indeed that Otters already have them present in their tissues. There is already proof that for one class of pharmaceutical drug, a bioaccumulation in Otters is indeed occurring.

A scientific study published in 2011 revealed that the commonly available and very widely used anti-inflammatory drugs ibuprofen and diclofenac were present in hair (fur) samples taken from 28 Otters that had been collected as carcasses from six counties across England. Both drugs were found to be present in the animals' hair, confirming that not only are Otters being exposed to these drugs, but they are also accumulating them in their tissues. We do not know what effect these drugs have on Otters, but one of them, diclofenac, has proved deadly for another group of animals elsewhere in the world: vultures.

The use of diclofenac as a veterinary drug in parts of Asia led to an almost total wipe-out of many species of vulture that live in this part of the world. The White-rumped Vulture suffered a 99% decline in numbers due to this drug, going from being the most numerous raptor species in the world in the 1980s, with a population of several millions, to one of the most endangered by the start of this century. Though they are now recovering, their numbers are still depressingly low, with a global population of just 6,000 individuals recorded in 2021. Birds and mammals are two different things, of course, and their bodily systems function in different ways. But that does not mean we should ignore the threat these drugs potentially pose for our river systems. After all, there are plenty of bird species, including such jewels as the Kingfisher, that live and feed on them.

Dogged persistence

Veterinary drugs in the UK are another source of pollution in our rivers, and in many cases we actively encourage these substances to go into the water. There has rightly been a lot of publicity in recent years about the use of a group of chemicals called neonicotinoids; their use in agriculture, especially on sugar beet crops, has caused much controversy. Neonicotinoids were banned for agricultural use in the UK in 2017 because of overwhelming evidence that they kill Honey Bees and other pollinators. Despite the ban, the UK Conservative government of the time authorised their use for four consecutive years in 2020, 2021, 2022 and 2023 under what was termed 'emergency authorisation'. The Labour government that came into power in 2024 stated that they would no longer allow the use of neonicotinoids, and in December of that year laid out their plan for the complete ban of three types – clothianidin, thiamethoxam and imidacloprid – that were used as agricultural pesticides and which are heavily linked to causing death and severe harm to bee populations. According to the government's plan, these neonicotinoids will no longer be legal to use for agricultural purposes in Britain. There will no longer be emergency authorisation for them.

But the devil is always in the detail. Neonicotinoids can still be legally used in Britain, and it is a use that does not need emergency authorisation. Their use is routine, and widespread throughout the country; you may even have them in your house right now. But this use is not as an agricultural pesticide. Instead they are used as a parasiticide veterinary drug to kill ticks and fleas, a drug applied to millions of pet dogs and cats every day, every month and every year.[15] One of the two neonicotinoids that are used on our pets for this purpose just happens to be imidacloprid, one of the neonicotinoids banned by the government for agricultural use because of the harm it does to bees and pollinators. As we have already seen, there are 13.5 million dogs in Britain today and many of these dogs are treated with these anti-tick and anti-flea parasiticides on a regular basis; indeed some of the products even recommend that your pet has monthly doses. There are 138 tick and flea treatment products currently on the UK market that contain imidacloprid. It is a product in widespread use in the pet sector, but it is one that has been banned, and banned with much fanfare, in the agricultural sector.

Scientific research has shown that once the product is applied to a dog (or a cat) it is absorbed by their body and subsequently can be found in their skin, hair, urine and faeces.

One single treatment for a large dog can contain enough imidacloprid to kill 25 million bees.[16] These very toxic chemicals are finding their way into our watercourses, and they are very effective insecticides, as the bees and pollinators have previously demonstrated. Their presence in our river systems threatens species like dragonflies and mayflies, insects whose larvae live as aquatic creatures. These insects and many more like them form the basis of the aquatic food chains on which, ultimately, our Otters depend.

The chemicals from these tick and flea treatments enter our rivers from a variety of sources, with the prime one being waste water. When dog or cat owners wash their pet's bedding, or even their own clothes with pet hair on them, they are washing the chemicals into the sewer. These chemicals then either go through the treatment works to be discharged as treated effluent back into our rivers, or they can be found in the sewage sludge that we use on our fields so wantonly and which then ends up in our rivers. But they can also be introduced into the rivers directly by dogs that are allowed, and often encouraged, to enter them by their owners. When dogs play in rivers and ponds, causing much delight for their owners, they can leave more than just happy memories behind.

Second helpings

Another mammal that regularly goes into the water is the Brown Rat, an introduced species in Britain and arguably our most numerous mammal.[17] It may be very numerous, but it is certainly not loved and as a consequence they are the target of large-scale poisoning both in the UK and across much of its now near-global range. The poisons mainly used to kill rats are known as anticoagulant rodenticides. The first part of the name describes accurately how the poison works – it stops blood from coagulating, or clotting, leading to the animal dying from internal haemorrhaging. The second part, though, is a bit looser in its descriptiveness: yes, it is a rodenticide, but it does not just kill rodents. It will also potentially kill most things that consume it, from hedgehogs to humans and owls to otters. They really should be referred to as biocides, substances that destroy living things.

Secondary rodenticide poisoning of predators is well known, with predatory bird species such as owls and kestrels and mammalian species such as foxes all being victims of our desire to be rid of rats. There has been little research done to see whether Otters are at risk from the use of rodenticides, but the research that has been conducted shows conclusively that they very much are.

A study in Germany, published in 2024,[18] tested the bodies of Otters found dead (primarily as road casualties[19]) over a 16-year period from 2005 to 2021. In total they tested the bodies of 122 animals. The results showed conclusively that Otters in Germany are exposed to these rodenticides; 87 of the 122 animals tested (just under three quarters of them) had anticoagulant rodenticides present in their livers. The authors of the study concluded that the Otters had acquired the poison via the food chain, and that elevated levels of rodenticide found in multiple Otters suggest that bioaccumulation of the toxin is happening up through the riverine food chain. Whilst this study was carried out in Germany, it does not mean that the same is not happening here in Britain. It is highly likely that it is, but we just have not looked for it yet.

There are numerous studies in the UK showing that other species of mustelid are in danger of secondary rodenticide poisoning, with Stoats and Weasels being at particular risk.[20] A Danish study published in 2018 showed that 99% of Stone Marten (a more terrestrial close relative of our own Pine Marten) livers tested positive for anticoagulant rodenticides, and 94% of Polecat livers tested also had them present.[21] Mustelids everywhere are exposed to these poisons that we routinely introduce into our environment. It should come as no surprise really that Otters too have these lethal poisons in their bodies.

From PFAs to microplastics, from ibuprofen to Ecstasy, from neonicotinoids to rodenticides, the list of potential pollution threats to our Otters and our riverine fauna seems endless.

The Future

A hundred years have passed since the publication of *Tarka the Otter* by Henry Williamson, a century of change for Otters, rivers and humans. But what of the next hundred years? On the face of it, the Otter is in a better position today than it was when Tarka first swam into the public consciousness. Their population is now most likely higher in the majority of the country, with the exception of the south-east of England, and they are no longer typically treated as vermin; although, as we have seen, there are still some that hold them in that light, the routine persecution they once faced is, thankfully, now a thing of the past. They are no longer hunted for our pleasure in a ritualised manner involving bright coats and hounds, and indeed their whole public image is now the polar opposite of what it was a century ago. Otters are widely considered to be cute and cuddly, soft toys of Otters are easy to find, and their image graces many a greetings card. They are used as branding, they are used as a marketing tool, and they are always a hit on any wildlife programme, as the producers of those programmes know only too well.

Habitat restoration on rivers and watersheds is happening – a truly positive action, although of course, more can and should be done. The historical damage caused by 'improvement' schemes is in some places being reversed, and with that reversal comes a growing appreciation of the need for us to have dynamic river systems, to maintain a natural balance in our watercourses that benefits us, as well as Otters, fish and other wildlife. The Beaver, a key species within any dynamic river system, has made a comeback, and this return has finally been granted legal approval in England following Scotland's approval several years ago. The Beaver's presence only improves things further for Otters. Things are looking rosy for the Otter, and the future can surely only bring better things.

Or can it?

The last hundred years of Otters in Britain nearly saw their complete demise; our liberal use of highly toxic chemicals in the countryside came very close to being their undoing. We should never forget that they did indeed become extinct in many parts of England, and we must not allow shifting base-line syndrome to obscure the fact that whilst they are now present again across the country, they have only very recently returned to every county in it. In some places their numbers are still very low, as their reclaiming of the rivers progresses at a very slow pace. Otters still have not fully recovered their lost ground.

Our river systems act as chemical conduits today, just as they did in the latter half of the last century. Whatever we put on the land or discharge in our waste water makes its way inevitably into our rivers. Whether this is tiny fragments of plastic leaching out of landfill or highly toxic human-created chemicals washing off agricultural fields, it all ends up in our rivers, our fish and our Otters. And this, for me and for many others in conservation, is the big worry.

Are forever chemicals, PFAs, the new organophosphates and organochlorines? Are they going to have the same impact as their predecessors, or are their effects going to be worse, more persistent and more complete?

We know that these toxic chemicals are already widespread throughout the environment, and are present in the tissues of fish and mammals; we know that Otters have been found with very high levels of them in their livers. Already in Britain our wildlife is loaded with PFAs. Is this load a burden that wild-life, and Otters specifically, can deal with? Or will that burden prove too much for them, leading to a collapse?

As I write this, I do not know the answer. Nobody does – we simply do not understand what this all-pervading presence of PFAs means for our wildlife. But it is a real worry and one concerning many people; more and more studies are being undertaken, more research is being done and all of them are revealing that Otters everywhere have these forever chemicals within them. And whilst this research is being done, whilst we are expressing concern, the flow of PFAs (and other chemicals) into our rivers and wider environment does not stop.

PFAs are chemicals that are potentially deadly, but they also have insidious sub-lethal effects. Much like the chemicals we poured so liberally into the environment in the mid-1950s, we have evidence showing that these substances can impair the breeding ability of animals that have the misfortune to come into contact with them, stopping them from reproducing suc-cessfully. If that happens, if the chemical load Otters in our rivers are subject to becomes too burdensome for them to con-tinue to breed, then a widespread collapse in their population is inevitable. The results of the fifth National Otter Survey of Wales were worrying, raising the spectre of potential popu-lation decline once more, but the results of the new national survey of England (the sixth) proved much more positive and seem to suggest that the earlier Welsh result was an anomaly. We can only hope that this is true.

The current picture may look promising in some ways, but the reality remains: our environment is flooded with PFAs and, as we have seen in the last chapter, many other very toxic chem-icals. We have not learnt the lessons of the past, we have chosen to ignore them instead. When an environment is flooded, the flood always flows eventually into the rivers and into contact with the Otters and other wildlife that live there. Can we really hope that this tide of persistent chemicals, chemicals already very much present in the tissues of Otters, will have no prejudicial effects on them and their lives?

There is a real lack of legislation surrounding PFAs in the UK, and this is compounded by the lack of legislation relating to them in sewage sludge. In America some farms have been forcibly closed because of contamination originating from the application of sewage sludge, and scientists there have come to a conclusion for one group of PFAs, the PFOs: if they occur at more than one part per billion in soil, then they will be absorbed by the crops growing on the land. Rates higher than that are the reason those farms in the USA were forced to close; it was considered too much of a risk for those crops to enter the human food market. But a recent study by the ENDS Report group has found that PFOs, despite being banned in the UK due to their strong links to cancer and human foetal development issues, were found at levels of 135 parts per billion in one soil sample from a farm in West Sussex, southern England, following the application of sewage sludge.[1] That is 135 times higher than what American scientists consider as the safe limit.[2] That farm has not been closed and is still producing food for human consumption.

Britain is too far behind when it comes to having stringent legislation in place to reduce the use of chemicals like PFAs, and the use of sewage sludge, which has been proved time and time again to be a key way that these forever chemicals (and a whole lot more) enter our environment. Other countries including those in the EU already have strong legislation concerning the manufacture, usage and disposal of PFAs as well as the use of sewage sludge, yet we do not. We chose to leave Europe and its legislation, we chose to take back control, but the sad reality is we don't have any real controls in place when it comes to these substances. Their pollution of our rivers is simply out of control.

There is, shockingly, hardly any legislative checks when it comes to using sewage sludge as a fertiliser on farm fields in the

UK. Yet we know that sewage sludge is the source of so much of the pollution in our rivers, including the above-mentioned PFAs. A newspaper article in March 2025 by the journalist and environmentalist George Monbiot looked at the subject in more detail.[3] In it he reports that other than a few scant tests for the heavy metals I mentioned previously, there is no testing at all for PFAs of any type, banned or not, and nor is there any testing for other potentially dangerous contaminants. A suppressed government report from 2017, however, showed that there were plenty of these contaminants likely to be present in the sewage sludge, including polycyclic aromatic compounds, dioxins, furans, benzopyrene, polychlorinated biphenyls (otherwise known as those everlasting PCBs) and phthalates. Most of that list are linked to cancers, reproductive issues and foetal development issues in humans. We are spreading this stuff on our food-producing fields and then eating the subsequent produce, and yet we have no idea of what we are really eating or what it will do to us in the long term.

But we are doing this through a choice of sorts, even if it is not our own individual choice. The contents of sewage sludge are well known by the very people who should be safeguarding our food production, but we choose as a society to continue to use it on our food-producing fields. Otters though do not have that choice, and nor does any other wildlife. They have to deal with the consequences of our choices, and we are making really terrible choices. Humans like to think that they are somehow different from other animals, but we are mammals at the end of the day; the way our bodies work is very, very similar to other mammal species like Otters. What is harmful for us is going to be harmful for other mammal species. If what we are spreading on our fields causes foetal development issues in humans, it is odds on that it can do the same in other mammal species too. The contents of the sewage sludge used in agriculture rapidly end up in our river systems and as we have seen with PFAs, they then end up in the bodies of our

Otters. In the future, in the immediate future, you would hope that we stop using sewage sludge as a fertiliser on our farm fields. If we are going to continue with it, then at the very least we must screen it for all potential pollutants and regulate its use accordingly.

When I began work on this book, the thought of having a national strategy for our river systems had never occurred to me. But the more I have researched, the more I have read and the more I have written, the more I have come to realise that this is desperately needed for a whole host of reasons, not least for the wildlife of those river systems. I am not alone in this epiphany.

Water UK is the trade association that represents all of the water companies and waste water companies in the UK. Founded in 1998, this umbrella organisation is seen by many as more of a problem than a possible solution to the issues affecting our rivers and their wildlife. After all, much of the blame for river pollution in this country is laid squarely at the door of the water companies and waste water companies, and we have already examined their views on the application of sewage sludge to our agricultural fields – it is something they seem to portray in a favourable light, despite the clear evidence to the contrary. But even though we differ greatly on that, Water UK do agree with me on the need to formulate a national strategy for our rivers.

On Water UK's website,[4] they outline ten recommendations they feel need to be implemented to create a better future for our watercourses. Number one on the list is the need to create a national plan for rivers. Water UK want this to be a clear, single, evidence-based, long-term plan agreed between the government, regulatory bodies, water companies, agriculture, highways and other sectors. That sounds good, but those other sectors must include those working for the needs of wildlife. Presumably this would be the statutory bodies that include wildlife in their remit, which in England would mean

Natural England; in Wales, Natural Resources Wales; and in Scotland, Nature Scot.

There are many tremendous wildlife NGOs in Britain doing great work and there should of course be involvement for them, but as there are so many of them, each with a slightly different agenda and each chasing the same pool of funding and members, this could be problematic. This highly fragmented collection of environmental charities can actually be a weakness, especially when drawing up such a wide-ranging and comprehensive strategy as a national rivers plan. Too often environmental NGOs can struggle to see the bigger picture, due to the blinkers of their own existence forcing them to focus on aspects that can help them access more funding and members. The conservationist and campaigner Mark Avery, in his book *Reflections*,[5] criticises our environmental NGOs, saying that many are too timid and they are not politically engaged enough. Ultimately, he says, this leads them to failing nature.

That assessment has been considered harsh by some, whilst others have wholeheartedly agreed. Personally, I do think there are too many organisations effectively doing the same thing; it does weaken the arguments when not all the NGOs decide to back a cause for reasons that do not always appear to be wildlife-related. Resisting the urge to go off on a tangent here, I will just say that if a plan for our rivers is to be drawn up – and it has to be, surely – then it needs to be drawn up quickly. Trying to involve a plethora of different environmental NGOs in this endeavour is simply not going to work; the process will very quickly get drawn out and bogged down. One wonders if that is the reason Water UK used the words 'other sectors', rather than specifically mentioning wildlife interests. Is the thought of engaging with these groups too onerous for them, perhaps? There are many farmers in the UK, but only really one organisation that represents them, the National Farmers' Union or NFU. There are many environmentalists and conservationists in the UK, but there are literally dozens and dozens of organisations that represent them. It becomes unworkable to deal with so many different groups. If Water UK want to discuss the needs of agriculture for the purpose of the plan, they deal with one organisation; but who do they deal with when it comes to the needs of wildlife?

The danger here is that they do not attempt to engage with any NGO, and instead deal with a statutory body. Unfortunately these bodies, despite being staffed by good people, can all too often have their position on issues weakened by their bosses in government. If a national river plan comes to fruition in the future, and it needs to be the near future in my opinion, the voice for wildlife has to be a strong and united one, or that voice risks being drowned out in a river fast flowing with other priorities.

Number two on Water UK's list of recommendations is the creation of a new 'Rivers Act', a single legislative act focusing on outcomes that, in the words of Water UK, have 'the flexibility to ensure that innovation, particularly in nature-based solutions, can be delivered'.[6] That sounds promising and much needed; one would assume that any such legislation would be part of the process required to deliver a national rivers plan, so perhaps recommendations one and two are interlinked.

The need for a national rivers strategy is abundantly clear, but it will not be easy to produce. There are many, many interests and perspectives to be incorporated, and the waters are sure to be muddied. But the need for it must be satisfied to safeguard our river systems and all that depend on them, be they human, animal, plant, or any other part of the natural world. Water UK's wording for their top two recommendations is a positive start, especially in their use of the phrase 'nature-based solutions'; these have to be key to any strategy, be it re-establishing Beavers or bends. I sincerely hope that in a hundred years' time, people are looking back at how successful the implementation of our national rivers plan has been.

One hundred years before I started to write this book on Otters, Henry Williamson was writing his own. I wonder whether he ever thought about what the world would be like in a century's time, how Otters and their rivers would be faring today in the 2020s compared to what he knew in the 1920s. If he did think about the future, perhaps he thought that things would stay pretty much the same when it came to his book's subject, that the public perception of the Otter would remain unchanged, that hunting them with hounds would still be a normal and socially acceptable pursuit, and that the rivers would just keep on flowing with life much like they always had. I of course do not know what he would have thought – I can only employ my own conjecture, and as I said at the very beginning of this book, his reported views and mine are very, very different.

But having said that, I am fairly certain that he would have been absolutely heartbroken at what we did to Otters (and other wildlife) in the latter half of the previous century, through our widespread use of chemicals.[7] He may have been an ardent supporter of the Otter hunt, but that doesn't mean he wanted Otters to face near-total extirpation from this country; after all, hunters want to have a quarry to hunt. But Williamson also clearly had an appreciation, a respect for and a love of Otters, and whilst I find it hard to translate that into hunting them, it comes across clearly in his own writings. Otters meant a lot to him. He would no doubt have been thrilled that the Otters have made a comeback from their nadir, that they are now doing well in his beloved Two Rivers Country once again, but I also imagine him feeling frustration and even anger that, even though Otters are now doing well there, and in the majority of places across the country, there is once again a chemical cloud looming large on their horizon. He, like me, would surely hope that the cloud does not develop into a storm.

I will not be around in a hundred years' time to know what the world will be like for Otters in the 2120s. I can only hope that they are still here, slipping smoothly into the waters of the Taw and every other river in the country. That they are still living their almost unseen existence and doing so in clean, safe waters full of healthy fish, in the vibrant, dynamic river systems that we all, both human and beast, deserve.

Journey's end

A loose, straggly – you could even call them ungainly – group of Jackdaws fly over the almost appropriately named Crow Point before they cross the open water, making their way towards where I sit. The brilliant blue sky and bright sunshine belie the reality of the cold March wind, a wind evidently strong enough to drive the line of towering white wind turbines further off on the western edge of the treeless high ground of Exmoor. The waters in front of me appear, at a brief glance, to be calm, but a closer look reveals quiet turmoil.

The large, brightly coloured buoy in the middle of these waters has repositioned itself, pointing now towards the out-of-sight town of Bideford a short way upstream. The tide has turned; the gravitational pull of the moon is now impelling the salty waters of the sea back towards the land, back into the mouth of the estuary. Crow Point marks the place where the Taw estuary meets the Torridge estuary, where Williamson's Two Rivers become one. But the two rivers do not just meet each other here; they also meet the sea, and with the tide now rising there are three channels of competing water, three differing flows, all coming together. Watching the motion of the low ridges of water in front of me, I soon become aware of the myriad directions in which they move. These are not concentric ripples or even lines of small waves, but a jumbled mess of converging and competing flows. The water's colour varies depending on which part of it I look at, but all are unfathomable to my colour-blind eyes, a *mezcla* (to use a Spanish word) of different tones, movements and no doubt stirred-up sediments.

The water rises fast in this estuarine coming-together. The wet muddy sand that has been visible since I got here disappears rapidly as I watch it; the three Brent Geese that had been loafing on it only minutes before do not appear to relish being forced up onto the low-lying rough rocks between it and where I am sat. Instead, they take to the wing to join others of their kind currently grouped on an equally fast-disappearing bar of sand out in the flow. Before long they all are forced to take flight, at least 23 of them, and head out seaward. It will not be too much longer before they are flying off on a longer journey, heading back to their breeding grounds much further north in the Arctic Circle. A Taw to tundra journey they have been making for centuries, millennia, a journey that once confused early medieval chroniclers into thinking that they, and the closely related Barnacle Goose, hatched not from eggs as other birds do, but from Barnacles themselves.[8]

A small group of Oystercatchers quicken their pace as they scurry along the soon to be inundated Bladderwrack-covered rocks below me. They are hoping to prise one last meal before being forced to move on. As I watch their black-and-white antics, a dark shape, only a few metres away from them out in the rising waters, catches my eye. I do not really see it properly, since my focus was on the Oystercatchers, but what I did see appeared to be a sleek black form slipping below the turbid surface. I instantly lose interest in the waders and scan the water, awaiting a resurfacing. My mind starts racing. Could it be? Could I be that lucky? The waters are broken by a long dark neck, followed by a low-slung body. The newly surfaced Cormorant takes a brief look around, before it slinks back below the water to propel itself along on its powerful feet as it seeks sustenance.

I give myself a wry smile and take a sip from my beer. Estuarine birdwatching is often a cold and lonely affair, but I am sat in a nice warm and friendly pub, perched on the edge of the confluence of the Taw and the Torridge and the sea, a pub with perfect large windows overlooking the waters and the life they hold. From my comfortable seat I look out at the end of the Taw. The river that started as a mere coalescence of drips on the high ground of North Dartmoor is now a broad estuary in North Devon. Its journey has come to an end.

My beer is very nice, though sadly it is not an Otter Ale, but I am sat in a Beaver. Or to be more accurate, The Beaver. This pub, perched on an old quay in the small town of Appledore, was so named as it was the place where Beaver skins from North America were unloaded to begin their journey to furriers elsewhere. The skins were imported in large numbers to satisfy our demand for the fur, a demand that had long before seen the demise of our own native Beavers. But today, the pub name is no more than a nod to the past, a nod done with a cheeky smile too no doubt. There are no Beaver skins here now, but as we have seen, there are once again living Beavers on the Taw. And of course there are Otters – not fictional ones, but real ones and in good number. However, today they are being their usual ever-elusive selves. I don't mind; sometimes just knowing they are there is good enough. I take another sip of beer and continue to watch the flow.

Species Mentioned in the Text

Sometimes the way we name species can be confusing. Species can be referred to by an informal common name, which is often an abbreviated form of its official common name – for example, in this book I use the word 'Otter' instead of the full-length official common name of 'Eurasian Otter'. But when mentioning another otter species, I have used that animal's full official common name to distinguish it from the Eurasian Otter.

For this guide to species mentioned, I have chosen to list each species by its everyday common name as generally used in the popular field guides. However, in some cases I have listed the species using the full official common name to avoid confusion between closely related species, American and European Mink being an example. The scientific name of each species is also listed to ensure absolute clarity!

Finally, in the text I have occasionally used a name that refers to a closely related group of species, for example 'willows', but I have not listed any willow species here as I have not referred to them as individual species in the text.

Mammals

Otter	*Lutra lutra*
American Mink	*Neogale vison*

Badger	*Meles meles*
Brown Bear	*Ursus arctos*
Brown Rat	*Rattus norvegicus*
Eurasian Beaver	*Castor fiber*
European Mink	*Mustela lutreola*
Field Vole	*Microtus agrestis*
Fox	*Vulpes vulpes*
Harbour Porpoise	*Phocoena phocoena*
Hedgehog	*Erinaceus europaeus*
Lion	*Panthera leo*
Lynx	*Lynx lynx*
North American Beaver	*Castor canadensis*
North American River Otter	*Lontra canadensis*
Pine Marten	*Martes martes*
Polecat	*Mustela putorius*
Rabbit	*Oryctolagus cuniculus*
Red Deer	*Cervus elaphus*
Ring-tailed Lemur	*Lemur catta*
Roe Deer	*Capreolus capreolus*
Sea Otter	*Enhydra lutris*
Stoat	*Mustela erminea*
Stone Marten	*Martes foina*
Water Vole	*Arvicola amphibius*
Weasel	*Mustela nivalis*
Western Spotted Skunk	*Spilogale gracilis*
Wolf	*Canis lupus*

Birds

Barnacle Goose	*Branta leucopsis*
Blackbird	*Turdus merula*
Blue Tit	*Cyanistes caeruleus*
Brent Goose	*Branta bernicla*
Bullfinch	*Pyrrhula pyrrhula*
Buzzard	*Buteo buteo*

Cattle Egret	*Ardea ibis*
Cetti's Warbler	*Cettia cetti*
Chiffchaff	*Phylloscopus collybita*
Collared Dove	*Streptopelia decaocto*
Cormorant	*Phalacrocorax carbo*
Corncrake	*Crex crex*
Dipper	*Cinclus cinclus*
Feral Pigeon / Rock Dove	*Columba livia*
Great White Egret	*Ardea alba*
Green Woodpecker	*Picus viridis*
Grey Heron	*Ardea cinerea*
Grey Wagtail	*Motacilla cinerea*
Jackdaw	*Coloeus monedula*
Kestrel	*Falco tinnunculus*
Kingfisher	*Alcedo atthis*
Little Egret	*Egretta garzetta*
Montagu's Harrier	*Circus pygargus*
Moorhen	*Gallinula chloropus*
Osprey	*Pandion haliaetus*
Oystercatcher	*Haematopus ostralegus*
Passenger Pigeon	*Ectopistes migratorius*
Peregrine	*Falco peregrinus*
Pheasant	*Phasianus colchicus*
Red Kite	*Milvus milvus*
Robin	*Erithacus rubecula*
Rook	*Corvus frugilegus*
Skylark	*Alauda arvensis*
Song Thrush	*Turdus philomelos*
Spanish Imperial Eagle	*Aquila adalberti*
Sparrowhawk	*Accipiter nisus*
Stonechat	*Saxicola rubicola*
Turtle Dove	*Streptopelia turtur*
Wheatear	*Oenanthe oenanthe*
White-rumped Vulture	*Gyps bengalensis*
Woodpigeon	*Columba palumbus*

| Wren | *Troglodytes troglodytes* |
| Zitting Cisticola | *Cisticola juncidis* |

Trees

Ash	*Fraxinus excelsior*
Atinian Elm	*Ulmus minor* 'Atinia'
Common Hawthorn	*Crataegus monogyna*
Douglas Fir	*Pseudotsuga menziesii*
Holly	*Ilex aquifolium*
Rowan	*Sorbus aucuparia*
Spindle	*Euonymus europaeus*
Wych Elm	*Ulmus glabra*

Other species

Amur Carp	*Cyprinus rubrofuscus*
Barbel	*Barbus barbus*
Black Bean Aphid	*Aphis fabae*
Brown Trout	*Salmo trutta*
Common Carp	*Cyprinus carpio*
Eel	*Anguilla anguilla*
Fox Tapeworm	*Echinococcus multilocularis*
Frog	*Rana temporaria*
Honey Bee	*Apis mellifera*
Salmon	*Salmo salar*
Signal Crayfish	*Pacifastacus leniusculus*
Tench	*Tinca tinca*
Three-spined Stickleback	*Gasterosteus aculeatus*
Toad	*Bufo bufo*
White-clawed Crayfish	*Austropotamobius pallipes*

The UK Wild Otter Trust

The UK Wild Otter Trust or UKWOT for short is a charitable trust founded and run by Dave Webb. It is based in the beautiful Taw valley in North Devon; as described in Chapter 10, the Trust exists to rescue wild Otters that are in immediate danger, rehabilitate them and then release them back into the wild where they belong. They do a tremendous job. Below is a look at the work involved for just one of the many Otters they help every year.

It was a hot summer's day in June when UKWOT received a call from a concerned member of the public, who had heard the plaintive cry of a young Otter cub and had soon located it, all on its own, alongside a busy road near the West Devon village of Broadwoodwidger. Due to the extreme heat of the day, and the very real chance of the young cub being run over, UKWOT asked the member of the public to pick up the young cub using a towel and to take it home so that it could be collected direct from them.

Young, distressed Otter cubs are fairly easy to handle – once the towel was placed around the youngster, it was easily transferred to a cat carrier box that the member of the public owned. With small cubs, the first 48 hours after they have been found is critical. The animals are usually extremely dehydrated and starving, and need rapid attention. UKWOT had collected the young animal and had it back at their rehabilitation centre within three hours of the initial phone call.

On arrival at the centre, the young Otter was immediately treated for dehydration; it was given fluids, antibiotics and some special home-made fish soup. The animal, a female, weighed in at 870 grams and was estimated to be between 12 and 14 weeks old. Following the fluids and feed the Otter cub was left alone to rest, after what had obviously been a very trying time for it.

On the following day, the team at UKWOT conducted a series of health checks on the young Otter, checking the animal's limb movements and the condition of its teeth, as well as assessing its body fat. Whilst doing this they also assessed the animal's overall behaviour. For this young Otter, which by now had been given the name of Widget after the location in which she had been found, all tests were passed satisfactorily. Widget now entered the rehabilitation programme.

For the first six weeks of this programme young Otters are kept inside in a purpose-built isolation unit, where they can be easily cared for and checked upon. This also acts as a quarantine station to ensure that the other Otters already in the rehabilitation programme are not at risk from any diseases a new arrival may be carrying. Widget responded well to the round-the-clock attention, and also to the food provided, reaching two kilograms of weight within those six weeks.

Once Widget had hit the two-kilogram weight mark, she was ready to be rehoused into the purpose-built outdoor enclosures. Widget's first enclosure was relatively small, ensuring that the team could easily monitor her progress, as well as enabling them to carry out daily weight checks to ensure that Widget continued to grow as she should. As Widget grew so did her appetite, and the daily food quantity she was given increased accordingly.

Widget progressed very well and was soon ready to be moved to a bigger enclosure, complete with a larger bath for her to enjoy and to get used to being in water. As the rehabilitation continued, so Widget was moved further along the line of enclosures, each one further away from the buildings and human activity. Even though Widget was helpless when she was initially rescued, she was a wild animal and if the rehabilitation was to be successful, it was vital that this was how Widget remained. Reducing the amount of human contact to an absolute minimum is a key part of the process.

Throughout the twelve months of rehabilitation Widget thrived and gained weight steadily. The team were happy with her behaviour and were confident that Widget was now ready to be returned to the wild. Widget was no longer a helpless cub weighing under a kilo. She was now a healthy young adult animal weighing 6.3 kg and, as Dave will attest, she was also very feisty.

Release sites are selected on how the team think the Otter will behave; normally a young adult Otter will have had plenty of experience of deep pools and fast-flowing water, but Otters that enter the centre as young cubs will obviously not have this familiarity, so sites have to be assessed and chosen with care. Widget was released at a site on private land in North Devon, which contained a slow-moving stream with plenty of natural food – a good place for Widget to re-enter the wild. All releases are carried out with support feeding and monitoring using trail cameras, so the team can check that the Otter has adapted well to being back in the wild. Widget took to her freedom like a duck (or should that be Otter?) to water. Thanks to the careful, diligent work of Dave and his team at UKWOT, they have never had to take a released Otter back in again. All have been successfully rehabilitated.

Widget was a success: from facing certain death as a helpless young cub, she was released as a thriving young adult ready to carve out a territory of her own. But this success comes at a financial cost, one that is entirely paid for by charitable donations to the Trust. The UKWOT team take no wages or expenses from the Trust, and nor do the many volunteers. All donations go towards helping rescue, rehabilitate and release the Otters.

Dave estimates that it costs around £3,500 per cub for the full 12-month rehabilitation process, with the largest of those costs being the food, electricity and water. The work that UKWOT carry out is fantastic and awe-inspiring. If you want to help them carry out this work, please see their website ukwildottertrust.org for details about how you can make a donation, or even adopt one of the young Otters they care for.

Acknowledgements

Writing this book has been a hugely enjoyable task, and one made much easier by the many people who have helped me along the way. Writing this sort of book is something you cannot do in isolation; I have chatted to many people about this book, sometimes asking questions, sometimes just talking over ideas. Thank you to everybody who I have spoken to.

Special thanks must go to Pete Burgess, who gave up his time (twice in one day on one occasion, thanks to my inability to understand time differences…) to talk to me about the fantastic work carried out on the River Otter in Devon as part of the River Otter Beaver Trial. I hope the other exciting nature recovery projects you are involved in are just as successful.

Huge thanks must go to Dave and Cathy Webb at the UK Wild Otter Trust for always making me feel so welcome when I visited. Thank you for providing me with lots of great information as well as the opportunity to see some of the Otters at close quarters, and of course thank you for the Scotch egg tip-off! But most of all, thank you for the tremendous work that you both do alongside your volunteers. The world needs more people like you.

Thank you to Peter Hawes and all at Bernard's Acre for a warm welcome on a very cold morning, and thanks also to Tim Penrose and his slightly bonkers but very brilliant team at Dartmoor Auctions. Thanks to Mark Darlaston for kindly providing me with information on Peregrines in Devon. I'll buy you a coffee the next time we meet on a ferry. Thanks also to Rupert Murray for many things, but especially for sending me the video of the Otter.

Thank you also to the staff at the Motor Museum Trust Research Service, the Road Lab team at Cardiff University and all involved in the Otter Project also at Cardiff University. Thank you to John Dawe for his hospitality and willingness to share his experience of Otters and fishing lakes; I enjoyed talking fences with you!

I would also like to thank Matt Larsen-Daw of the Mammal Society for kindly taking time to chat to me about the unreleased data from the latest Otter survey of England – much appreciated. This brings me on to the subject of the organisations, conservationists, researchers and campaigners that do so much for Otters and rivers in Britain: thank you all.

I would also like to thank Nigel Massen and all at Pelagic for helping me develop this from an idea in my head to a finished book. Thank you. My thanks also to Sara Magness who had the task of copy-editing my manuscript – thank you for your diligence and patience. They say don't judge a book by its cover, well the cover is brilliant! Thank you to Rachel Hudson for her fantastic illustration.

Final thanks must go to Jo for all the support and love you give me.

Notes

Introduction

1.   Rachel Carson, *Silent Spring* (Penguin, 1962).

Chapter 1: Bubbling Beginnings

1.   Laverock is an old, alternative name for the Skylark.
2.   But of course, this is not *wilderness*, even though it is one of the remotest places in southern England in terms of distance from roads and houses. The landscape of Dartmoor has long been heavily shaped by human activity. And sheep, far too many sheep.
3.   This is the East Dart, the second of the Dartmoor Dart rivers, the other, the West Dart springs forth a few miles to the south before meeting and joining the East at the appropriately named Dartmeet.
4.   The Wolverine, the largest member of the Mustelidae, looks more bear-like than weasel-like, but it too has an elongated body and relatively short legs; it also possesses an incredibly strong bite.
5.   The others being Weasel, Stoat, Polecat, Pine Marten and Badger. The American Mink is here too, but as the name suggests, it is an introduced species.
6.   There have been records of Rabbit remains in archaeological sites that pre-date the Norman Conquest, but their presence is almost certainly as a result of the Rabbit's burrowing activities, which can take them down through layers of history.
7.   Rabbits were very numerous in the first half of the twentieth century, but when Myxomatosis was deliberately introduced to Britain in 1953 it had a devastating impact on their numbers, perhaps wiping out as much as 95% of their population in just a couple of years. The drastic reduction of a potentially important food source for the Otter could have had an effect on them, but the mid-1950s delivered a far more drastic blow to Otters, as will be seen in a later chapter.
8.   Seals diverged from the mustelids approximately 50 million years ago.
9.   Healthy body fat percentage in humans varies somewhat with sex and age.
10.   Roger Deakin, *Waterlog* (Vintage Books, 2000), p. 185. The book is primarily about wild swimming around Britain, but it is so much more than that and is, in my opinion, a must-read.
11.   Ted Hughes read the address at the beginning of Williamson's memorial service in 1977; the poem quoted has the appropriate title of 'An Otter'.

12. An Old English word that was originally used to describe small wooded areas, its use as a name for an Otter's den is first recorded in 1590.
13. Paul Chanin, *Otters* (Whittet Books, 1993), p. 14 – an excellent volume.

Chapter 2: Us and Them

1. The character of John Ridd lived on Exmoor. He described himself as being both of Devon and of Somerset; I claim him for Devon here purely to fit the narrative and therefore apologise to any Somters.
2. Gavin Maxwell, *Ring of Bright Water* (Longmans, 1960).
3. The service, previously known by the more prosaic North Devon Line, was renamed the Tarka Line in 2001 when it was operated by Wessex Trains. It first opened in 1854.

Chapter 3: Poisoned Arteries

1. It was thankfully pointed out to the government that not only was the Spindle not the only host for that generation of aphids, but that the work of Professor A. C. Hardy had already shown that clouds of the minute insect drift in from the continent on air currents every year in a form of aerial plankton. Even if the British government had succeeded in wiping out the Spindle, we would still have had aphids on our beans. And as it happens, the Dig for Victory campaign was a complete success even with aphids.
2. Data sourced from Felicity McWilliams, 'Equine Machines: Horses and Tractors on British Farms 1920 to 1970', PhD thesis, King's College London, 2020.
3. Domestic pigeons are a different species altogether from the Woodpigeon, and have a variety of names from racing pigeons to feral pigeons; they are all actually a species called the Rock Dove. Their domestication goes back for at least 5,000 years and initially was to provide food rather than to send messages.
4. During the Second World War, Britain used around a quarter of a million domestic pigeons for communication purposes. There was even a Pigeon Section at the Air Ministry.
5. Some nine months after the war in Europe had ended.
6. A sad phrase that is applied to a population or species that although still existing has lost the means to reproduce. Probably the most famous example is Martha, a female Passenger Pigeon that was the last of her kind. Even though she lived for many years in an American zoo, she was alone and unable to breed; the species was functionally extinct even though Martha still lived.
7. Otter hunting with hounds was formally made illegal by the Wildlife and Countryside Act of 1981.
8. M. Darlaston, 'A Review of Devon's Peregrine Population', *Devon Birds* 73 (2), 2020: 3–10.
9. Much of the river system in the city of Bradford has been built over and covered up, much like many of London's rivers, such as the Fleet.

10.   The fire risk was real, with documented records showing that on at least one occasion the canal was aflame. Hydrogen sulphide gas bubbling up from the coal sludge at the bottom of the waterway was the reason for its flammability.
11.   See Ian D. Rotherham, 'The Urban Wild: A Perspective from Northern England', *British Wildlife* 36 (2), 2024: 89–96 for details of the rivers of Sheffield, their surroundings and how they have come back.

Chapter 4: The Return

1.   PCBs or polychlorinated biphenyl compounds were commonly used in the manufacture of a wide variety of products including paint, waterproofing, paper, lubricating oils, plastics, etc. Their various usages started to be banned in the UK from 1981 but, as with many fabricated chemical compounds, they are extremely persistent pollutants.

Chapter 5: Lost, Found and Returned

1.   In *Tarka*, Williamson used the term Landrail, one of many old names for this bird.
2.   W. S. M. D'Urban and Rev. Murray A. Matthew, *Birds of Devon* (R. H. Porter, 1892).
3.   It should be noted that in keeping with the trend of the time, by encountered they meant shot; as if the birds didn't have enough problems!
4.   There has been an ongoing attempt at reintroduction of Corncrake in East Anglia, but whether it will lead to a viable self-sustaining population is debatable.
5.   The resident Heron in Tarka's life was a bird called Old Nog.
6.   Little Egret data taken from an article by Ian Parsons in the August 2019 issue of *Bird Watching* magazine.
7.   Our warming world is behind a northward range expansion of many species of wetland birds. As some wetland sites in the south of Europe become unsuitable due to lack of water, the birds are seeking sites further north.
8.   George Montagu (1753–1815) was a pioneering ornithologist writing definitive books on the subject. He is best known today for having the Montagu's Harrier named after him. George Montagu, *Ornithological Dictionary* (J. White, 1802), quoted in W. S. M. D'Urban, *Birds of Devon* (R. H. Porter, 1892), p. 171.
9.   The four other names before the Taw are also major Devon rivers.
10.   Often rendered by those that like to put birdsong into words as 'What's my name? Cetti, Cetti, Cetti.'
11.   Ironically, Signal Crayfish were introduced to many parts of Europe because they are resistant to the plague; what was not realised at the time is that they also carry it and are important vectors for it.
12.   There is a species of mink that is native to Europe, the European Mink. Sadly it is now listed as being critically endangered. Although sharing the

name of Mink with the American species, it is actually more closely related to our own Polecat. European Mink have never occurred in Britain.

13. Yes, you read that right. Lions are native to, and were found, in southern Europe and were reported as being common in Greece in 480 BCE. It was of course humans who extirpated them from the continent by 100 CE.

14. The English National Otter Surveys are carried out by the Mammal Society. The 2024 survey data is used with kind permission of the Mammal Society, as it is unpublished at the time this book goes to press. For further details of the surveys please see Chapter 11.

15. A recent article in *British Wildlife* magazine suggests that eradication is now within reach. Vince Lea, 'A Mink-free Britain Is Now within Reach', *British Wildlife* 35 (5), April 2024: 313–17.

16. Roger Lovegrove, *Silent Fields: The Long Decline of a Nation's Wildlife* (Oxford University Press, 2007).

17. See Dr Hugh Webster, 'No Place for Lynx?', in *Great Misconceptions: Rewilding Myths and Misunderstandings*, ed. by Ian Parsons (Whittles Publishing, 2024).

18. For a good look at the Wolf's demise see Jim Crumley, *The Last Wolf* (Birlinn Limited, 2010).

19. The book was written by Bartholomaeus Anglicus in *c.*1245 before being translated into English in *c.*1398.

20. In more recent times castoreum has been 'milked' from live captive animals, but back when Beavers were wild natives in Britain, the animals were killed to obtain it.

21. The practice of releasing Beavers without licence or agreements has been given the name of Beaver bombing. It is illegal.

22. Another is that the released Beavers are of the correct species, the Eurasian Beaver. Elsewhere in Europe the North American Beaver has been released into the wild, either deliberately or via escapes; they are not the same species.

23. Although not primary hosts, humans can also be infected by the tapeworm which then proceeds to cause serious illness.

24. This illustrates just how reckless any unofficial release of a Beaver is – yes, Beavers should be here and it is great that they are back, but there were many powerful interest groups that didn't want them to return. Imagine the hue and cry if the Beavers of the River Otter had brought with them the pernicious tapeworm and that parasite had spread into the domestic dog population, resulting in people's pets being put down. If this had happened the entire official Beaver reintroduction project would likely have been heavily tarnished by the resulting outcry, and most probably it would have been halted.

25. As are many upland rivers too.

26. The Scottish government approved their release in Scotland in 2008.

27. The picture, described as one of the nation's greatest and most popular paintings, now hangs in the National Gallery in London.

28. Pronounced like 'Which', the tree becomes both a question and its answer: Which Elm is native? Wych Elm is native.

29. Born *c.*4 CE.

30. As poor a name as English Elm, Dutch Elm Disease was named because it was Dutch scientists that first discovered it; but the strain of pathogen

that wreaked so much destruction to the elms of Europe in the 1960s and onwards arrived here from North America on imported timber, and is thought to originate from Japan.

31. Quote taken from the Forest Research website: https://www.forestresearch. gov.uk/tools-and-resources/fthr/pest-and-disease-resources/dutch-elm-disease-ophiostoma-novo-ulmi/dutch-elm-disease-history-of-the-disease/

32. The bark beetles that feed on the inner bark of Elm, and unwittingly spread the fungus as they do so, only feed on stems that are greater than 10 cm in diameter. Young suckers and small-diameter growth are safe from the beetle and therefore the fungus.

Chapter 6: A Disturbing Enemy?

1. The percentage of Devonians living in the two cities and the urban area of Torbay is 44%.

2. Dog licences were introduced to Britain in 1867, but they didn't become mandatory until 1959. They were abolished in 1988.

3. The earliest record of a domesticated dog was found in a grave alongside the skeletons of a man and a woman in what is now modern-day Germany; the grave was dated as being 14,200 years old.

4. I can attest to their adroitness at kicking, having received one in the face when trying to intervene in a dog–deer dispute. It hurt. A lot.

5. Two good examples: Michael A. Weston and Theodore Stankowich, 'Dogs as Agents of Disturbance', in *Free-Ranging Dogs and Wildlife Conservation*, ed. by Matthew E. Gompper (Oxford University Press, 2013) https:// wilderness-society.org/wp-content/uploads/2019/04/Dogs-as-agents-of-disturbance-Michael-A.-Weston-and-Theodore-Stankowich.pdf; Alex Lees, 'Gone to the Dogs', *BirdGuides*, 30 January 2022, https://www.birdguides. com/articles/conservation/gone-to-the-dogs/

6. A quote often attributed wrongly to Ratty in the book *Wind in the Willows* by Kenneth Grahame – that character actually said 'messing about in boats'. Messing about on the river was the title of a song sung by Josh MacRae, a top-20 hit in 1961. The song was written by Tony Hatch, who also co-wrote the theme tune to *Neighbours*… Otters need good neighbours too.

7. Don't get me wrong, having a dog can be a fantastic thing that can bring many benefits to us humans, but that doesn't mean that we should turn a blind eye to the impacts those dogs can have on wildlife and the environment.

Chapter 7: Why Did the Otter Cross the Road?

1. There are over 8,000 miles of public road in Devon, making it the largest county road network in all of Britain.

2. Data taken from the National Motor Museum Trust Research Service. Private cars made up over 800,000 of the total; goods vehicles, taxis and

agricultural vehicles made up the rest. From an Otter's point of view the type of vehicle that kills you doesn't matter.

3.   The study carried out by The Road Lab at Cardiff University recorded a total of 89,000 mammal casualties during the ten-year period. Otters made up 2% of that number.

4.   This map is publicly available on the internet and can be found at https://www.cardiff.ac.uk/otter-project/research/map. It is somewhat sobering to look at.

5.   The installation of the wildlife shelf received a lot of press coverage in August 2024 – see for example Christian Fuller, 'Shelf Will Help Otters Safely Cross Road', *BBC News*, 18 August 2024, www.bbc.co.uk/news/articles/cd052z5zz5no

Chapter 8: Fishing for Answers?

1.   Often referred to as stew ponds.

2.   The book is still in print!

3.   Environment Agency, *A Survey of Freshwater Angling in England*, 2018. An online copy can be found here: https://www.gov.uk/government/publications/a-survey-of-freshwater-angling-in-england

4.   Simon Scott, 'The Truth about… Otters!', *Carpology*, 13 November 2017, https://www.carpology.net/article/features/the-truth-about-otters-/

5.   There are plenty of other examples of how the AIF has helped fisheries resolve conflicts with Otters; see for example The Angling Trust, 'The Angling Improvement Fund – Funding Support for Predation Defences', 8 April 2024, https://anglingtrust.net/2024/04/08/the-angling-improvement-fund-funding-support-for-predation-defences/

6.   In 2022 a study estimated that pet cats in Britain kill between 160 million and 270 million wild animals and birds a year. Tara J. Pirie, Rebecca L. Thomas and Mark D. E. Fellowes, 'Pet Cats (*Felis catus*) from Urban Boundaries Use Different Habitats, Have Larger Home Ranges and Kill More Prey than Cats from the Suburbs', *Landscape and Urban Planning* 220, April 2022, https://doi.org/10.1016/j.landurbplan.2021.104338

7.   I use the word 'fishermen' because from a look through the comments, they seem to have been posted by males.

8.   The Salmonidae is the family to which both Salmon and Trout belong.

9.   But there are other types of pollution that can, as we will see in a couple of chapters' time…

Chapter 10: Rescue, Rehabilitation and Release

1.   Douglas Firs are most likely the tallest trees in the world – it's just that we felled the bigger and taller specimens, leaving the lesser-valued (timber-wise) Coast Sequoia to hold that particular epithet for now. Their tallest specimens can be as high as 106 metres, but Douglas Fir were much taller, with specimens over 130 metres recorded before they succumbed to the saw.

2.   See Appendix 2 for further details of the work of UKWOT; it is a great charity completely run by volunteers, and the appendix gives details on how you can make a donation if you so wish.

3.   Details of the letter and its content: The Wildlife Trusts, 'Charities Challenge Ministers: Fix the Planning Bill, or Nature Will Pay the Price', 9 April 2025, https://www.wildlifetrusts.org/news/charities-challenge-ministers-fix-planning-bill-or-nature-will-pay-price

Chapter 11: Lessons Learnt or Forever Failing?

1.   Numerous studies have looked at organophosphate poisoning in humans, at acute poisoning and its immediate effects. Studies such as this one, however, have also shown that long-term usage can lead to lasting neurological problems: R. Stephens et al., 'Neuropsychological Effects of Long-Term Exposure to Organophosphates in Sheep Dip', *The Lancet* 345 (8958), May 1995, https://doi.org/10.1016/S0140-6736(95)90976-1

2.   The map can be found here: https://watershedinvestigations.com/home/find-out-whats-polluting-your-local-rivers-lakes-and-coast/. It has various layers that you can switch on and off, enabling you to see just what is in your local watercourse. It is a tremendous piece of work, but be warned, once you have used it you will not be able to view your local river in the same way ever again.

3.   Taken from an article in the *Guardian* newspaper: Rachel Salvidge and Leana Hosea, 'Otters among UK Wildlife Carrying Toxic "Forever Chemicals", Analysis Shows', *Guardian*, 17 January 2025, https://www.theguardian.com/environment/2025/jan/17/otters-among-uk-wildlife-carrying-toxic-forever-chemicals-analysis-shows

4.   In case you are wondering what happens to the remaining 13%: 6% is used in land reclamation, 4% of it is incinerated and the remaining 3% is used mainly as a fuel for cement production. Data on sewage sludge, or biosolids, taken from the independent scientific research organisation the James Hutton Institute.

5.   Data also taken from the *Guardian* article previously cited. The fish sampled and recorded on The Watershed Pollution Map all had levels of PFAs in them higher than the level the EU is thinking of setting as the environmental standard for them.

6.   A single hair from your head measures on average 80,000 nanometres in width.

7.   James D. O'Connor et al., 'Microplastics in Eurasian Otter (*Lutra lutra*) Spraints and Their Potential as a Biomonitoring Tool in Freshwater Systems', *Ecosphere* 13 (7), 2022. https://doi.org/10.1002/ecs2.3955

8.   U. Nopp-Mayr et al., 'Microplastic Loads in Eurasian Otter (*Lutra lutra*) Feces – Targeting a Standardized Protocol and First Results from an Alpine Stream, the River Inn', *Environmental Monitoring and Assessment* 196 (707), 2024, https://doi.org/10.1007/s10661-024-12791-z

9.   D. Harley-Nyang et al., 'Investigation and Analysis of Microplastics in Sewage Sludge and Biosolids: A Case Study from One Wastewater Treatment Works in the UK', *Science of the Total Environment* 823, 2022.

The study can be seen in full here: https://ore.exeter.ac.uk/repository/handle/10871/128986

10. A summary of the study by the James Hutton Institute can be read here: James Hutton Institute, 'New Study Finds 1,450% Increase in Microplastic Levels within Soil after Four Years of Sewage Sludge Application', 13 March 2025, https://www.hutton.ac.uk/new-study-finds-1450-increase-in-microplastic-levels-within-soil-after-four-years-of-sewage-sludge-application/

11. The EIA have produced a report called *Cultivating Plastic*, the relevant part four of which can be accessed here: https://eia-international.org/wp-content/uploads/2023-EIA-UK-Cultivating-Plastic-Part4-Caution-and-Alternatives-SINGLE-PAGES.pdf

12. Data obtained from various sources including the National Archives: https://www.ons.gov.uk/peoplepopulationandcommunity/crimeandjustice/articles/drugmisuseinenglandandwales/yearendingmarch2024

13. A report by the Chartered Institute of Environmental Health on the study: Nicola Smith, 'Rivers in National Parks Are Polluted with Pharmaceuticals', Chartered Institute of Environmental Health, 5 September 2024, https://www.cieh.org/ehn/environmental-protection/2024/september/rivers-in-national-parks-are-polluted-with-pharmaceuticals

14. *Great UK WaterBlitz: September Report*, September 2024. The full report can be read here: https://earthwatch.org.uk/wp-content/uploads/2024/10/Great-UK-WaterBlitz-report-Sept24_WEB.pdf

15. Pesticides kill pests, parasiticides kill parasites. Semantics are important when it comes to government rules and press releases.

16. The data I have quoted on the use of neonicotinoids in pet treatments is taken from the Imperial College London's briefing note on the subject: Hayley Dunning, 'Toxic Pet Flea and Tick Treatments Are Polluting UK Freshwaters', *Imperial College London*, 20 March 2023, https://www.imperial.ac.uk/news/243875/toxic-flea-tick-treatments-polluting-uk/

17. Some reports state that there are perhaps 80 million Brown Rats in the UK; others say there are 10 million, quite a disparity. If we take the 80 million figure then they outnumber the Field Vole, which is also reported as being the most numerous mammal in the UK with a population estimated at 75 million individuals. Whatever the actual number, there are a lot of Brown Rats in Britain.

18. J. Regnery et al., 'First Evidence of Widespread Anticoagulant Rodenticide Exposure of the Eurasian Otter (*Lutra lutra*) in Germany', *Science of the Total Environment* 907, 2024. An abstract can be seen here: https://doi.org/10.1016/j.scitotenv.2023.167938

19. But what caused them to be road casualties? Animals suffering from anticoagulant poisoning don't die immediately – before they do their behaviour alters, as they become far more lethargic and less aware of their surroundings. With road casualties, just because the actual cause of death is blunt force trauma caused by a collision with a vehicle, we must not forget to consider exactly why that animal was hit.

20. For example: R. A. McDonald et al., 'Anticoagulant Rodenticides in Stoats (*Mustela erminea*) and Weasels (*Mustela nivalis*) in England', *Environmental Pollution* 103 (1), October 1998, https://doi.org/10.1016/S0269-7491(98)00141-9

21.   Morten Elmeros et al., 'Exposure of Stone Marten (*Martes foina*) and Polecat (*Mustela putorius*) to Anticoagulant Rodenticides: Effects of Regulatory Restrictions of Rodenticide Use', *Science of the Total Environment* 612, January 2018, https://doi.org/10.1016/j.scitotenv.2017.09.034

Chapter 12: The Future

1.   The UK's number-one source of intelligence for environmental professionals, according to their website.

2.   Tess Colley, 'Water Companies Selling Sludge Fertiliser Containing Banned "Forever Chemical" to Farmers', *Ends Report*, 20 September 2024, https://www.endsreport.com/article/1889418/water-companies-selling-sludge-fertiliser-containing-banned-forever-chemical-farmers

3.   George Monbiot, 'What's in the Millions of Tonnes of Sludge Spread on to UK Farmland? Toxic Waste – and Ministers Don't Care', *Guardian*, 21 March 2025, https://www.theguardian.com/commentisfree/2025/mar/21/tonnes-sludge-uk-farmland-sewage-farms-toxins-food-system

4.   www.water.org.uk

5.   Mark Avery, *Reflections: What Wildlife Needs and How to Provide It* (Pelagic, 2023).

6.   See https://www.water.org.uk/protecting-environment/protecting-rivers-and-coasts, under 'What needs to be done', '2. Protection in law'.

7.   Henry Williamson died in 1977 at the age of 81, the very year that it was proved that the pesticides introduced in the 1950s had caused the catastrophic decline in the Otter population.

8.   With no comprehension of migration, the seemingly obvious solution to these chroniclers as to why they never saw the birds nesting or why they never witnessed young goslings was that the birds hatched from the small crustaceans that live on the rocks where the birds spent much of their (wintering) time. This theory was given credence by the English herbalist John Gerard, who claimed to have witnessed them hatching from the said barnacles on several occasions. One has to wonder exactly what herbs he was growing.

Also available from Pelagic Publishing

Reflections: What Wildlife Needs and How to Provide It,
Mark Avery

Clinging to the Edge: A Year in the Life of a Little Tern Colony,
Richard Boon

*Treewilding: Our Past, Present and Future Relationship with
Forests,* Jake M. Robinson

*The Atlas of Early Modern Wildlife: Britain and Ireland
between the Middle Ages and the Industrial Revolution,*
Lee Raye

*Bats & Ladders: Stories from the Real World of Bat
Conservation,* George Bemment

Fenland Nature, Duncan Poyser and Simon Stirrup

Low-Carbon Birding, edited by Javier Caletrío

The Hen Harrier's Year, Ian Carter and Dan Powell

*Crossbills and Conifers: One Million Years of Adaptation and
Coevolution,* Craig W. Benkman

*No Island Too Far: Searching for Seabirds on Remote Specks of
Land,* Michael Brooke

A Wildlife Guide to Georgia, Brecht De Meulenaer

Purposeful Birdwatching: Getting to Know Birds Better,
Rob Hume

*Wildlife Photography Fieldcraft: How to Find and Photograph
UK Wildlife,* Susan Young

*Rhythms of Nature: Wildlife and Wild Places Between the
Moors,* Ian Carter

Birds & Flowers: An Intimate 50 Million Year Relationship,
Jeff Ollerton

Wild Mull: A Natural History of the Island and its People,
Stephen Littlewood and Martin Jones

pelagicpublishing.com